LASER EXPERIMENTER'S HANDBOOK—2ND EDITION

DELTON T. HORN

TAB BOOKS Inc.
Blue Ridge Summit, PA

SECOND EDITION
FIRST PRINTING

First edition copyright© 1979 by TAB BOOKS Inc.
Copyright © 1988 by TAB BOOKS Inc.
Printed in the United States of America

Library of Congress Cataloging-in-Publication Data

Horn, Delton, T.
    Laser experimenter's handbook / by Delton T. Horn. —2nd ed.
       p. cm.
    Includes index.
    ISBN 0-8306-9115-4 ISBN 0-8306-3115-1 (pbk.)
    1. Lasers. 2. Lasers—Experiments. I. Title.
TA1675.H67 1988   88-25038
621.36'6—dc19   CIP

TAB BOOKS Inc. offers software for
sale. For information and a catalog,
please contact TAB Software Department,
Blue Ridge Summit, PA 17294-0850.

Questions regarding the content of this book
should be addressed to:

Reader Inquiry Branch
TAB BOOKS Inc.
Blue Ridge Summit, PA 17294-0214

Cover photograph courtesy of Coherent, Inc., Palo Alto, CA.

Contents

Introduction

YOU DO NOT NEED A MASTER'S DEGREE IN PHYSICS TO EXPERIMENT WITH lasers and related devices. This book is intended to give a theoretical and practical starting point for the electronics hobbyist who wants to get into this fascinating high-tech field.

I have, by necessity, drawn certain preliminary assumptions in respect to the reader's background in the general sciences. A science background is an important prerequisite for any explorer in this complex field of study.

Without a doubt, this subject field, which deals with the inseparable aspects of atomic nature, molecular construction, and electromagnetic wave propagation, is perhaps one of the most complex and technically profound of all the branches of science and physics.

The first half of this book deals with the theoretical basics of the laser. Every attempt has been to make this theoretical material as painless as possible, but this field is simply not for anyone who hates science and math.

While a firm theoretical background is given, a number of simple projects are presented to get the reader started in experimentation. Because this is a book for beginners, a conscious decision was made to keep the projects as simple as possible. Only very low power laser devices are used to minimize the risks as much as possible, but always remember that even a low-power laser is potentially dangerous if it is used incorrectly.

Laser experimentation is not recommended for the casual hobbyist. If care is not taken in constructing an ordinary electronics project, the odds are it just will not work. If care is not taken with a laser, serious injuries could result to the hobbyist or others.

I do not want to scare anyone off, but this is a complex subject, using potentially dangerous equipment. Put forth the necessary effort and use the necessary precautions, and lasers will be a very fascinating and rewarding extension of your electronics hobby.

Chapter 1

The Laser:
Yesterday,
Today, and Tomorrow

LASERS ARE A FAIRLY RECENT INVENTION. THEY STILL SEEM VERY FUTUR-
istic, but they are not quite as new as most people assume. On a
theoretical level, lasers were first conceived back in 1917 as an
outgrowth of Einstein's atomic theories. Einstein mathematically proved
that under special conditions, controlled radiation could be obtained
from an atom. Such controlled radiation is at the heart of any laser
device.

HISTORY

Work on practical lasers got started in the 1950s. The earliest
laserlike devices were actually masers. The term *maser* is actually an
acronym for *microwave amplification by stimulated emission of
radiation*. A laser is a specialized type of maser. The word *laser* is
an acronym for *light amplification by stimulated emission of radiation*.

The first practical maser was developed in 1954 by Dr. Charles
H. Townes of Bell Laboratories. This invention brought him the coveted
Nobel Prize in 1964.

By 1958, Townes, along with Dr. Arthur L. Schawlow, had advanced the propositions of how light can be made coherent by the
amplification of stimulated emission of radiation. The new device they

were describing was originally called an optical maser, but it eventually came to be known as the laser.

In 1960, Dr. Theodore H. Maiman, who was then working at the Hughes Aircraft Company in Malibu, California, had succeeded in actually applying the theories to a practical device. Working with a synthetic ruby, he was rewarded by a very minute beam of coherent red light. This was the first such beam of laser light to be propagated by the laser process. Maiman's ruby laser was soon followed by many additional practical laser devices using various materials.

From these humble and seemingly infantile roots came a marvel that is perhaps surpassed by no other basic invention since the wheel. A great many applications have been found for the laser, and new ones are constantly cropping up. During the printing of this book, it is probable that even newer uses will be found for the laser.

INSIDE A LASER

Maiman built the first prototype laser from a small, man-made ruby rod that had been doped with chromium. This rod was about 1 centimeter in diameter and 2 centimeters long, carefully machined to optical tolerances. The two ends were cut precisely parallel with each other, and the ends were then polished and silvered to enhance reflection. Light within the ruby rod bounced back and forth between the two mirrorlike ends. The rod served as sort of a resonant chamber. One of the ends was made slightly less reflective than the other. A small opening in the silvering provided an escape path for the coherent beam of light, which is the output of the laser.

The rod was mounted in a wrap-around flash tube filled with xenon gas. This flash tube was quite similar to those used in high-speed photography. In operation, a capacitor was discharged through the flash tube, causing it to bathe the ruby rod briefly in a brilliant burst of light. When this happened, a substantial amount (but not all) of the light was absorbed into the ruby rod. This actuated the chromium atoms the ruby had been doped with.

Once the chromium atoms had been energized by the externally pumped light, they became elevated to a higher energy level (sort of like being kicked by an external body). Earlier theory had predicted that the chromium atoms could not stay at this unnaturally higher energy value for very long. The high energy state could be held for about 1 millionth of a second. After this brief time, the chromium atoms dropped back or decayed, giving up their shortly held energy to the surrounding ruby atoms, thus activating them and causing lasing.

When the ruby atoms lase, the photons of energy start surging back and forth from end to end between the confines of the silvered resonator ends. They do this until enough of the atoms have developed the required energy to cross the threshold of the mirror resistance. The atoms break through, and emission occurs. The output is a very narrow, very powerful coherent beam of red light. The red color is a result of characteristics of the ruby crystal. Other lasing materials might produce beams of other colors.

To continue the lasing action, the flash tube must be repeatedly fired at a regular rate. This type of laser is called a *pulsed-wave laser* because of the repeated firing. Some later devices are continuous-wave lasers. The basic operating principles are similar.

In addition to ruby, many materials—both gaseous and solid state—have been used for lasers.

LASERS OF ALL SIZES

If you think the original prototype laser was small for its ability, then consider this: Bell Laboratories has already manufactured a fully functional laser that is actually smaller than a grain of salt. Of course, its power is accordingly diminished, but it is a functional laser nevertheless, which is remarkable enough (Fig. 1-1).

Of course, larger lasers also have been built. Gas lasers tend to be larger than solid-state lasers. Some high power gas lasers have been built to lengths in excess of 200 feet (Fig. 1-2). The maximum output power of these gas lasers is directly proportional to the length between the mirrored ends (this length is called the *resonance length*).

Fig. 1-1. Minute lasers are capable of tremendous power.

Fig. 1-2. A scientist exhibits a C amplifier—one of the largest in the laser chain under development.
Photograph courtesy of Lawrence Livermore Laboratories.

These devices range from the carbon dioxide laser, which can vaporize diamonds and all other substances known to humanity, to the gentle surgical laser, which is being used to perform delicate eye surgery while the patient is fully awake.

The power ranges of lasers begin at the miniscule level, which even a mosquito could not feel. The other extreme of laser power is represented by a model made by Westinghouse. Without even focusing, the output of this device is more than 750 trillion watts (W) of power. This amount of power is analogous to all of the water going across Niagara Falls being shot in one instant squirt through a water pistol.

PRESENT APPLICATIONS

The doors of industry that have been opened by the laser are virtually countless (Fig. 1-3), and new doors are being opened almost

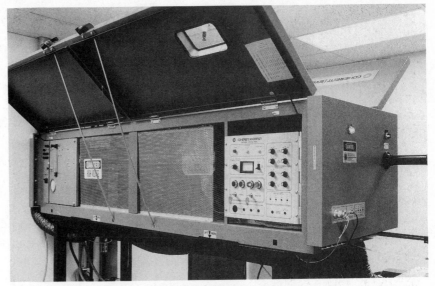

Fig. 1-3. An industrial production laser.
Photograph courtesy of Coherent Laser Division.

every day. Metals and elements that were once thought to be indestructible can now be literally sent up in a cloud of bluish green vapor at the concentrated thermal touch of a laser.

The laser has a number of obvious applications in weaponry, cutting, and welding. Less obvious applications are plentiful in such diverse areas as photography, medicine, communications, and even entertainment. Just a few of these applications are presented in the following sections.

Holography

Holography is a specialized form of photography that uses a laser. The prefix *holo* means whole. A holograph is a very complete photograph of an object. A laser can be used to take or project a very lifelike, three-dimensional picture of any object. In a hologram (a picture taken by the holograph process), you can see the entire picture in depth—even around the corners and opposite sides of objects. The effect is quite incredible.

The idea of three-dimensional photography is not particularly new, but practical holography has been around only for a couple of decades. The reason that the practice of holography was developed only recently is that the process depends upon the division of light rays in a simultaneous and exact manner, which was not possible with ordinary light sources.

Actually, the image you see in a hologram is the result of the re-formation of the reflected image at a distance from the lighted object. The reflected rays are reconstructed at the point of junction wherever the mirrors are aimed.

Researchers are working on creating holographic movies and television. Practical commercial devices of this type are not likely to appear on the market soon, but there is no theoretical reason they cannot be developed in the future.

In an application somewhat related to holography, the laser is being shrewdly employed in some museums to detect forgeries of valuable paintings and sculptures. In this application, the laser produces a very tiny, controlled beam of high-intensity light that vaporizes a minute particle of the item being tested. (The damaged spot is so small that it is not visible to the unaided eye.) The gas produced can be viewed with a spectrograph, and the elemental properties of the material can be scrutinized. The laser used for this type of operation is known as a *microprobe*.

Medicine

At first glance, the power of the laser might seem to be primarily destructive. However, the beam can be focused to an extremely fine point, and it can be very useful in performing surgery that might be too delicate for even the sharpest scalpel. One of the first areas of medical treatment where the laser proved its worth is optical surgery (operations performed on the eye).

In optical surgery, a precisely controlled laser beam is used to perform operations that only a few years ago were considered almost impossible. The laser device used by optometric surgeons is called a *photocoagulator.* It can literally weld the retina back in place for those thousands of people yearly who would otherwise suffer blindness from a detached retina. The photocoagulator makes use of the lens of the patient's eye to focalize the light energy to just enough intensity to fuse the delicate membrane of the retina. This operation is very safe and almost always successful, thanks to the laser. This operation can permit a patient to have the retina welded and be able to ride home in an hour. This is about the same period of inconvenience as having a tooth pulled.

The laser also finds its place in dentistry, where the teeth, after losing their hard, glossy shell (enamel), are irradiated by the laser. It has been conclusively proven through tests that teeth treated in this manner definitely are superior to untreated teeth in the retardation of decay.

A fine-beam laser can also be used to burn away tiny blood clots in veins. The clots, if untreated, could lead to heart attacks or other serious ailments.

The *biolaser* is another medical application of the fundamental laser. It directs pinpoint thermal energy in absolutely controlled dosages. It is used primarily in destroying cancerous cells on the microscopic level.

The beam in the biolaser is kept to a diameter not exceeding 40 millionths of an inch. The biolaser procedure has been perfected to such an extent that even chromosomes can be split at their nucleus. (The problem now is in determining which chromosome to split.)

Bloodless surgery is another significant development that supports the use of lasers in the operating room. Because of the concentrated heat at the very point of cutting, the focalized beam automatically cauterizes the tissue as it cuts, thus closing off all arteries and capillaries as it progresses.

Biolasers are being used successfully in the removal of tumors and skin scars with very little noticeable sign of the removal. This is obviously very desirable in cosmetic surgery, for example, where the removal of a birthmark might otherwise leave a scar that is worse than the original birthmark. This type of treatment also has proved to be quite effective in the removal of tattoos, which were once assumed to be permanent.

You might well ask how it feels to have a biolaser beam touch your skin. Most people would probably assume that it would be extremely painful. Actually, it is no more painful than having a small bit of melted wax drop on your skin. It is more discomfort than pain. Biolaser surgery is better than traditional surgery because often no anesthetic is required, and there is generally no after-sting or soreness, which might be experienced after traditional surgery. This is because the laser affects a very small portion of the area being treated.

Communications and Entertainment

A beam of light can be modulated and encoded in a manner similar to a radio wave (Fig. 1-4). With ordinary light, this generally is not very practical because information density remains low. However, a tight, coherent beam of laser light is an ideal carrier for almost any type of encoded data, either analog or digital.

Modulated lasers are being employed increasingly in long-distance telecommunications. The laser light beam is transmitted over a special cable made of fiberoptic strands. A fiberoptic cable comprises a bundle of individual strands. Each strand is thinner than a human

Fig. 1-4. A researcher communicates via the laser communicator receiver and the Metrologic neon laser.
Photograph courtesy of Coherent Laser Division.

hair, and each strand can carry a beam of light in a manner similar to the way an ordinary wire carries an electrical signal. The internal construction of the optic fibers permits the light beam to follow any desired path, even around curves and sharp bends. Ordinary light can be carried by optical fibers, but laser light, being more coherent and powerful, can travel farther without intermediate amplification. Modulated laser light is also less prone to transmission errors that might occur if any ordinary light source were used.

Laser-based devices are showing up in more and more homes. Compact disc players, like the one shown in Fig. 1-5, use a laser to detect encoded digital data on a small, silvered disc. A typical compact disc is also shown in Fig. 1-5. Figure 1-6 shows a simplified cross section of a compact disc, magnified many times. Digital data is encoded on the disc as a series of pits and bumps in the disc surface. The disc is backed with a reflective material. A clear coating is placed over the encoded surface to protect the disc from dirt and damage. In operation, a beam of laser light is reflected off the disc surface. This light is reflected back to a detector, which can determine—by the amount of light reflected and its angle—whether the beam is currently focused on a bit or a bump. A laser beam is required for

Fig. 1-5. The ADS CD3 compact disc player with a typical compact disc.

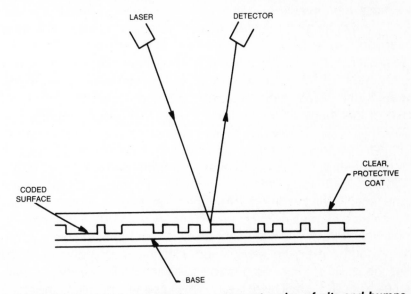

Fig. 1-6. Digital data is encoded on a compact series of pits and bumps.

this application because an ordinary light source could not be focused finely enough on a single spot on the disc. Of course, a very small, low-powered laser is used in a compact disc player.

Video disc players work in much the same way, except the recorded data is encoded in an analog code, rather than in a digital code.

These are just a few of the current applications of lasers. There are a great many others.

THE FUTURE

Within the narrow confines of a specific technology it often is not too difficult to make a few reasonable predictions based on past developments and current research. Most of these predictions are likely to be fulfilled in the near future.

Obviously all the possibilities cannot be covered here because there is a lot of research being done in laser applications. A few of the most important and most interesting areas that have high potential for future development are discussed in the following sections.

Radio Physics

It seems that radio, television, and communications have breathlessly awaited the advent of the laser. The laser has created so many new opportunities in the area of telecommunications that it is almost unbelievable.

One example that seems to be from science fiction is telestar reception. This technology allows scientists to hear the bursts of light and radiation waves from stars that emitted them over 1 million years ago. Scientists can also hear radio waves that were presumedly created long before even dinosaurs roamed the earth. You might suspect that there would be little to hear from space. Actually, the heavens are so full of *star talk* (electromagnetic waves from stars) that a maser must be discretely and carefully tuned to the particular frequency sought. The bleed-over is enormous because all of the millions of stars are *talking* (creating electromagnetic waves) at the same time. Perhaps someday (maybe soon) people might receive— and be able to transmit in return—some very definite intelligence to one of those stars. The star might be so far out from our solar system that it cannot even be seen, even with the most powerful of all telescopes. But it might be possible to hear it all the same. (Or maybe not. Nobody know if there is any other intelligent life in the universe.)

By using a laser beam for transmitting communications (using fiberoptics as discussed earlier in this chapter), it might be possible to transmit millions of telephone conversations simultaneously on the same beam. Current systems scarcely scratch the surface of the theoretical possibilities.

Similarly, thousands more television programs might be transmitted. This increase would be possible because of the inherent stability of the laser bandwidth of emission that can be provided by a carrier wave of, for example, 10^{14} hertz (Hz) of a beam.

Military Applications

A few decades ago, pulp science fiction literature was full of predictions of *death rays,* which were more or less handheld laser weapons. While this application has not really been developed, the military has put the laser to several more sophisticated uses, and considerable research is being done on future military applications of the laser.

The laser finds unequalled use in radar detection. Military lasers are scanning the hemisphere in alertness for anything that appears to be suspicious or lethal to the nation's protection and security.

Lasers are also used as blind-vision detectors for directing airplanes and missiles.

The best known, and most controversial military application of the laser is the Star Wars program. The *Star Wars* plan would place a number of satellites in permanent geo-synchronous orbit around the Earth. Any incoming missiles could be detected, targeted, and destroyed above the atmosphere. In theory, this is a purely defensive system; however, it would not be difficult to modify the system to target ground-level sites. In addition, such a defensive system could indirectly give a nation first-strike capability. The attacked nation would be unable to fire back in defense after an attack. As often happens, the technological problems are much easier to solve than the political problems.

Industrial Applications

The laser finds its greatest applications in machining, welding, and cutting operations. Materials that were once thought to be unusable because of their resistance to machining are now rendered almost pliable to the intensity and purity of the focalized laser beam. There is no known substance that cannot be worked by the laser.

Pieces as small as radio-tube filaments and components can now be precision-welded while still inside of their vacuum enclosure. This can be done without even removing the tube from the circuit. This fusion welding is accomplished by the beam passing through the glass walls of the tube and focalizing at a convergence point inside the tube. Thus, expensive electronic tubes that otherwise would have required replacement can be salvaged.

With system guidance for machine processes, precision can be assured to the millionths of an inch on lathe operations. Precision can be as close as ½ inch on tunnel excavation work where tunnels several miles long might be begun at the opposite sides of a mountain. The meeting point of the two partial tunnels will be as close as plus or minus ½ inch. Such precision would be unimaginable without a laser.

Interferometers have been used to detect changes in length, such as in bridge spans, to 1 billionth of an inch!

The conventional *gyroscope,* used for directional stability, has now met stiff competition that the mechanical gyroscope cannot compete with. The new laser gyroscopes are communicating changes in the Earth's orbital path that were never realized before. The mechanical gyroscope is dependent upon the Earth's gravitational influence, which can limit it in certain applications. The laser gyroscope is an electronic device, and its operation is not affected by gravity.

In the past, the process of drilling small holes in diamonds for wire-drawing dies was a two-day job. The work was tedious and quite expensive. A pulse of laser energy can do the job just as well (if not better) in a matter of a few minutes.

The precision of the laser in manufacturing is best shown by its use in *trimming* carbon-type resistors. Minute quantities must be trimmed (removed) to correct the resistor ohmic value. The quantity to be removed might be as small as 1 billionth of a gram. The amount is so infinitesimal that its removal can be verified only by electrical units so small that another laser must be used to prove the measurement.

Other Applications

Experiments have been conducted to develop a system in which blind people could be equipped with a laser transceiver. The system would enable the blind to detect the presence of objects and determine what direction to take to avoid contact with the objects. It is hoped that this system might also be able to provide the user with a one-direction outline or scan of the object detected. This would, in some respects, restore a type of vision to the blind.

Electrical power might be transmitted by laser beams, but only with an appreciable loss of efficiency. When the problem of power loss is eventually overcome, expect to witness the disappearance of all the electric poles and transmission lines from the countryside and the replacement of them with almost unnoticeable antennae for transmission of laser beams.

LASER MATERIALS

Since the appearance of the very first ruby rod type of crystal laser in 1960, many additional types of lasers have been developed, and they derive their lasing ability from diverse properties. Theoretically, any substance can be made to lase under the proper conditions. For many substances, the required conditions might be impractical to achieve. Even so, the future possibilities are almost limitless. The only comment that can be made safely is that there is no definite end point to the future of this sophisticated electronic marvel. It is so simple, yet so profound.

The earliest practical lasers used a solid-state crystal as the lasing substance. Soon after, a number of gas lasers appeared. The latest development is a new type of semiconductor laser called the *injection laser.* The injection laser is a special semiconductor diode that bears many similarities to the LED (light-emitting diode). This device will be discussed in more detail later in this book, and it will be used in the projects.

The working laser is relatively new, but the potential for the laser has existed ever since the first crystal formed on the earth. Unknowingly, a primitive gas near-laser was assembled as early as the late 1800s, when gas-filled Geiger (radioactivity sensitive) tubes were first used in experiments. How close scientists were then to creating the catalyst that would someday affect the lives of us all! The technology existed at the time to construct a practical laser, if only the idea had occurred to someone. Only people's cerebral inertia kept them from reaching out to the unknown, dreaming things that are not, and asking: "Why not?"

No energies can be spared in the continued development and research of this natural instrument of the universe. Pray that advancement will lead not to destruction, but to the civilization of dreams.

Chapter 2

Matter, Energy, and Atomic Nature

TO UNDERSTAND HOW THE LASER WORKS, YOU NEED TO KNOW SOMETHING about the nature of matter and how atoms are put together. You do not need a degree in atomic physics, but the subject must be touched upon. The discussion is kept as simple as possible. This chapter explores the basic principles by which lasers and masers function.

Ordinarily you might think of matter and energy as two distinct and very different things. Matter is a physical object, which can be readily detected and measured. You can touch it, and you can see it. Energy, on the other hand, is more elusive. You cannot directly touch or see energy. You can detect or measure its effects on matter, but this sensing is indirect.

Despite these fundamental differences, current atomic theory holds that matter and energy are one and the same. They are simply two differing manifestations of the same basic phenomena.

Matter is a physical representation of energy. Conversely, energy can be defined as an end product of matter or as an extended result of the essence of matter.

Undoubtedly you have seen Einstein's famous equation, $E = mc^2$, even if you do not understand all of its implications. This equation defines the equality of energy and matter. In the equation,

E represents energy, *m* represents mass (matter), and *c* represents the speed of light. Spelling these items out, the equation is: "Energy equals mass times the speed of light squared." Obviously, even a very tiny piece of matter is equivalent to a large amount of energy.

Neither matter nor energy can be created or destroyed, although one can be converted into the other. This is a fundamental precept of physics. If anyone ever disproves the law of conservation of matter and energy, the entire science of physics will fall apart.

If a given mass were completely reduced, or disintegrated, an equivalent quantity of energy would remain in its place. The equivalent would be a quantity—rather than a weight or mass—because energy does not have mass. The principle of the atomic bomb is a very good example: a given quantity of matter is converted almost instantly into its equivalent energy. A small atomic bomb will produce an explosion equal to that of hundreds of tons of ordinary explosives. If the conversion is made a little slower, and under more controlled conditions, the result is nuclear power. A few pounds of uranium can produce the power equivalent of several tons of coals. (Incidentally, an atomic, or nuclear, power plant cannot possibly explode as an atomic bomb explodes. Completely different processes are involved, and such an explosion would be a physical impossibility.)

You have learned that matter is but one facet of energy, and thus, energy is but one facet of matter. From this you could conclude that everything in the universe is a product of matter and energy. The total *product,* or resultant, of all the mass and of all the instantaneous prevalent energy equals a mathematical constant at any one given time. Think of it: in the entire expanse of this universe, there is a constant, which when applied mathematically, accounts for every gramweight of matter multiplied by all of the existing energy at any particular instant! The makeup of the total (that is, how much matter is in the total compared to how much energy is in the total) may vary, but the total cannot be changed.

You have concluded that matter and energy are essentially the same, but separate them again for convenience in discussing their individual properties. The discussion will start with an examination of the more physical composition of matter. Later in this book, the counterpart of matter, energy, is discussed.

ATOMS

All matter, be it any item, substance, compound, or thing, is made up of molecules. A *molecule* is the smallest particle of a nonelemental substance that still retains the characteristics of that substance.

Molecules are in turn made up of atoms. An *atom* is the smallest particle of an element that still retains the characteristics of that element. (A molecule may include atoms of different elements.)

Atoms are made up of still smaller particles: mainly protons, neutrons, and electrons. These particles are identical for all atoms, regardless of the specific element. The only difference, on the subatomic scale, between different elements is the number of these particles contained within the atom.

Protons are particles contained in the *nucleus,* which is a clump of particles at the center of the atom. Most of the mass of the atom is contained in the nucleus. Each proton has a small positive electrical charge.

Neutrons are also contained in the nucleus. They are similar to protons, except they are electrically neutral (no charge, either positive or negative).

Electrons are contained in rings surrounding the nucleus; the rings are like orbits of the planets around a star. Electrons are much smaller and have much less mass than protons or neutrons. Each electron has a small negative charge, which is exactly equal to the electrical charge of a proton. The electrical charge of the electron has the opposite polarity of the charge of the proton.

Normally, the number of electrons in an atom is exactly equal to the number of protons. The net electrical charge of the ordinary atom is zero. Under certain conditions, some electrons can be taken away from an atom, or extra electrons can be forced into orbit around the nucleus. In either case, the atom is then known as an *ion*, and it has an electrical charge. If there are too many electrons, the atom becomes a negative ion. If there are too few electrons, the atom becomes a positive ion.

The basic structure of an atom is illustrated in Fig. 2-1. This drawing is greatly simplified, but it gives you the general idea.

A Look through the Microscope

To visualize atomic structure more clearly, imagine looking through a super microscope. The substance you will see is just common water, or H_2O. Figure 2-2 illustrates several views at different amounts of magnification.

In Fig. 2-2A, a simple drop of water is shown as it appears to the unaided eye. By increasing the magnification of the microscope, you will be able to see that this drop of water is made of a great many individual molecules.

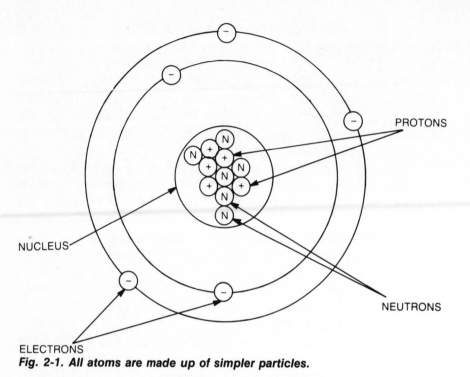

PROTONS

NUCLEUS

NEUTRONS

ELECTRONS

Fig. 2-1. All atoms are made up of simpler particles.

Next, increase the magnification even more to focus on a single molecule, as shown in Fig. 2-2B.

You can see that this molecule of water is actually composed of three parts: a center part of an oxygen atom and two atoms of hydrogen, which are attached to the oxygen atom in a Mickey Mouse ear fashion.

After moving the magnification up one more notch, observe that what appeared to be three items are now something like 13 much smaller units, as shown in Fig. 2-2C. These units are three nuclei (the nuclei of two hydrogen atoms and the nucleus of one oxygen atom) and 10 electrons (eight electrons belonging to the oxygen atom and one electron belonging to each of the hydrogen atoms).

In Fig. 2-2D one of the hydrogen atoms is isolated. Notice that it is composed of a planetary arrangement of a nucleus with only one *nucleon* (nuclear particle), in this case a proton, and one electron orbiting about its outer periphery. Notice that an ordinary hydrogen atom has no neutrons.

Next refocus the microscope on the oxygen atom to see it in all its splendor. This atom has a nucleus with 16 particles: eight protons and eight neutrons, as shown in Fig. 2-2E. Since there must be a matching electron for each and every proton, count eight electrons in orbit around the oxygen nucleus.

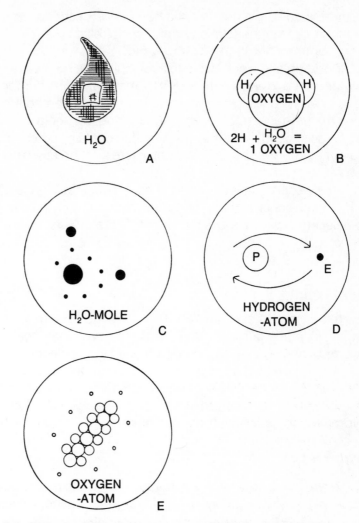

Fig. 2-2. A single drop of water as seen through a microscope.

All that in one little drop of water! Bear in mind that the smallest visible drop of water will contain literally thousands of individual molecules.

Without delving into chemistry any more than really needed, quickly refresh your memory by recalling that the outermost ring of the electron flight path or orbital arrangement serves as the combining or hooking facility by which the individual atoms hook onto other atoms to form molecules. Atoms combine to form the thousands of different compounds and molecules of which matter is made. The ability of any atom to unite with another atom to form a molecule is known as its *valence*. The outermost electron orbit is called the *valence ring*.

Valence is nothing more than the number of electrons less than or more than eight in the total outer number of electrons. Eight is a very important number in the atomic world.

As an example, oxygen has a total of eight electrons separated into two rings. Each individual ring has the maximum number of electrons it can hold. Rings farther from the nucleus can hold more electrons than the innermost ring. The first ring can hold only two electrons. The second ring of the oxygen atom contains six electrons. The outermost ring is the only one of interest in determining the valence.

Only the electrons in the outermost ring affect the valence of the atom. Remember that eight is the critical number. Oxygen has six valence electrons, which is less than eight. Therefore, the valence of oxygen is:

$$6 - 8 = -2$$

Oxygen has a valence of minus two. This gives the oxygen atom the ability to pick up one or two extra electrons from a *donor-type* atom, such as hydrogen. Because hydrogen has only one electron to donate, two hydrogen atoms are needed to provide the two electrons that the oxygen atom needs to bring its outermost ring up to a total of eight electrons. This is why the formula for water is H_2O: there are two hydrogen atoms for each oxygen atom.

Energy

You have had quite a microscopic introduction into matter. A very important part was deliberately left out for a good reason. Next, you will be reintroduced to the molecule, but in a different manner. The part left out of the discussion so far is energy. (Recall the equality of energy and matter discussed earlier in the chapter.)

It is *energy* that must be added to or taken away from atoms. The energy must be added externally or released internally, depending on the specific process or compound in question.

As an example, study the proton in Fig. 2-3. It is identical to any other proton that could be detached from any other atom of any other molecule, regardless of the individual type or character of matter it originally represented. The single proton in Fig. 2-3A is not special in any way.

All protons, neutrons, and electrons are identical and equal in both qualitative and quantitative dimensions. The difference in the type or character of matter is governed solely by the number of individual

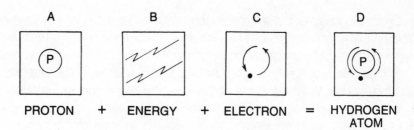

A   B   C   D

PROTON + ENERGY + ELECTRON = HYDROGEN ATOM

Fig. 2-3. The formation of a hydrogen atom.

particles, and the geometrical manner in which they are arranged, on both the atomic and molecular scale. A proton is a proton, a neutron is a neutron, and an electron is an electron.

Having established that important issue, take the proton (Fig. 2-3A) and add a certain quantity of energy to it (Fig. 2-3B). Bring these two together with one captured electron (Fig. 2-3C), and you have a hydrogen atom, complete with the necessary components: matter and energy (Fig. 2-3D).

The hydrogen atom was selected for this example because of its inherent simplicity; it is the simplest possible atom. It has the least possible number of electrons (one) and the most basic nuclear structure: one proton and zero neutrons.

It must be emphasized that all atoms of all elements are essentially similar in their multiple forms. They all have a nucleus of at least one proton and one or more neutrons. The nucleus is orbited by a number of electrons equal to the number of protons in the nucleus. This is true for all normal (non-ion) atoms ranging from the simple hydrogen atom up to and including the uranium atom, which contains 92 protons, 143 neutrons, and 92 electrons for a total atomic weight of 235.

All atoms, from the lightest to the heaviest, have *mass.* Mass of atomic particles is expressed by *atomic weight,* due to the respective quantity of protons and neutrons. Each proton and neutron has an atomic weight of one. The mass of the electron is so small that it is disregarded. For the record, the mass of an electron is approximately $\frac{1}{1847}$ the mass of either a proton or a neutron, It would take 1,847 electrons to equal the mass of a single proton or neutron, so disregarding the mass of an election is justified.

In some cases, there are variant forms of certain elements called *isotopes*. An isotope has a different number of neutrons than is normal for that type of atom. Isotopes seldom occur in nature in any quantity, but they can be artificially synthesized. Most isotopes have more neutrons than normal, so the atomic weight is higher than that of the natural form of the element. For example, if a neutron is forced into the nucleus of a hydrogen atom, the atomic weight jumps from the normal value of one up to two. When water molecules are formed with this type of hydrogen atom the result is known as *heavy water*.

In addition to mass, an atom also contains energy. To be specific, an atom contains two types of energy. One is the energy that is housed in the nucleus, which holds the protons and neutrons in place. This type of energy is known as *binding* or *packing* energy; binding energy is what *nuclear energy* is all about.

The second type of energy within the atom is *electron energy,* which has less potential and is easier to deal with and exploit than nuclear energy.

Nuclear energy deals with energy levels and values that require very elaborate devices—such as Van de Graaff generators, cyclotrons, and betarons—either to extract the energy of the atom or to rearrange its construction for other purposes.

It is important to remember that nuclear energy is an inescapable part of nature. It is contained within every single atom in the universe. Most people do not realize it, but everything is radioactive; some elements are just more radioactive than others. Uranium, for example, is more radioactive than ordinary oxygen, but ordinary oxygen is radioactive too.

For this discussion, attention will be limited to the more accessible electron energy. Electron energy exists in several forms. This energy is the frequency at which the electron is vibrating, or the distance from the nucleus to electron in its orbiting flight (or in what energy level or *shell* the electron resides). These factors affect the energy level by either *vibrational frequency* (in Hertz) or *shell occupation*. It requires more energy for the electron to move farther way (or to occupy an outermost shell or level) than it requires for the electron to remain close to its influence-exerting nuclear counterpart (proton). The position that is closer to the nucleus is considered the *home shell* of the electron. The *electron level* is that level the electron will occupy in its *natural state* or its *normal excitation level*. This energy level is sometimes called *ground level* or the *unexcited state.*

The normal excitation level is that state of being (or energy level) at which nature causes (with no outside interference) the molecule or atom to occur. Some external force is required to break the atom or molecule out of the normal excitation level. The level of energy in the unexcited state is affected by several factors, including the molecular temperature and the presence of any electronic or electromagnetic forces within the molecular environment.

Thus, a change within an atom or molecule could be effected by altering either the temperature or electrical state. If either of these conditions were changed, the total value of the molecular or atomic level or state of energy would be changed.

Consider the effects of temperature first. The temperature of a substance is a direct indication of its energy on an atomic or molecular level.

As heat is added to a substance, its temperature rises. On the microlevel, the speed or frequency of the vibrating molecules is increased. The higher the temperature of this substance is raised, the more energy it receives. Eventually—before the substances melts or vaporizes—a point is reached where, because of the unaccustomed new level of *thermal* excitement, the molecules become very *vibrant*. They start to glow, having reached a condition known as *incandescence.* (This is the principle behind the incandescent light bulb.) The point of incandescence is caused by the changes in the atomic energy level; the electrons are so vibrant and energetic that they fly farther away from their normal orbital paths. They have absorbed the energy to do so from the external source of additive energy, which in this case is heat. (Remember, energy can be neither created nor destroyed.) The electron is now a greater distance from the nucleus than it would be under normal, unexcited temperatures or circumstances.

For this electron to return back to its normal (and closer) orbital shell, it must give up that extra energy. At every shell or home orbit, there is only a certain energy content that an electron residing in that orbit can possess. Any additional energy must be disposed of. This disposal of excess energy is accomplished by ejecting or emitting the acquired energy. The electron can then return to its normal home shell.

The energy emitted by the electron is in the form of a *packet*, or small quantity of energy, called a *quanta of energy.* This quanta of energy has been emitted on such a frequency as to place it in the visible light range of the *electromagnetic wave spectrum.* That means that the waves at this frequency can be detected visibly. Due to this *photo effect*, the packet of energy can be described as a *photon* of energy.

This very superficially covers the *quanta theory of matter*, which is advanced physics. Fortunately, you do not need to be too concerned with more than the basics here.

The energy level also can be affected by applied electrical energy or light energy. (Light is part of the electromagnetic spectrum.) This leads to the conclusion that the energy level of an atom and its ultimate molecule can be affected by an externally applied energy.

In Fig. 2-4 there are three environments containing four atoms. The atoms are depicted as they usually exist in their normal, ground-

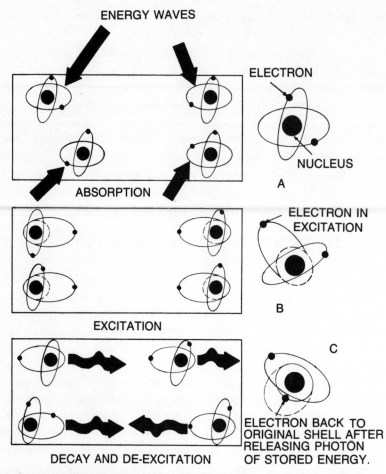

ENERGY WAVES

ELECTRON

NUCLEUS

A

ABSORPTION

ELECTRON IN EXCITATION

B

EXCITATION

C

DECAY AND DE-EXCITATION

ELECTRON BACK TO ORIGINAL SHELL AFTER RELEASING PHOTON OF STORED ENERGY.

Fig. 2-4. Atoms and electrons go through absorption, excitation, and decay in three separate environments.

level state in Fig. 2-4A. Arrows of energy are directed, or *applied*, into these ground-level atoms.

Figure 2-4B shows how the atoms appear after absorption of the energy (arrows) that hit them. You can see how the individual atom appears after becoming excited by absorbing the applied energy in Fig. 2-4B at the right. The electron has absorbed this energy and now orbits in a shell of greater magnitude than the shell it occupies in the unexcited state. Thus, the electron and its orbital position have been expanded in order to accommodate this higher level of energy.

The excited atom and its far-flying electron, as it now exists after releasing the momentarily absorbed energy, is shown in Fig. 2-4C. Notice the arrows of energy packets leaving the atoms. The individual electron has now returned to its closer or lower-level shell, only after giving up or radiating the applied energy that caused it to become excited in the first place.

ATOMIC EXCITEMENT

All atoms of matter normally exist at a certain energy level (ground level). At least the greatest majority or *population* of the number of total atoms of a given mass of matter exists at this *population level*.

By applying energy to this population, the total of atoms can be excited by way of *electron shell expansion.* They then go from their original ground level to a new level of excitement. The majority of atoms exist at a ground level of normally low excitement. The relatively small percentage of the population that exists at any higher level is due to the raising of the energy level. The addition of energy causes the population of atoms to be *inverted* from the normal state to a new excitement level. Therefore, the expression *population inversion* indicates that the energy has inverted the population of atoms from the normally occurring low percentage of randomly excited atoms, to a new, higher percentage of the population in the excited or inverted state.

You have witnessed that the atoms in Fig. 2-4C return to their former ground level after having once absorbed the applied energy (having become excited). This reversion back to ground level is known as *decay*, which is a good description of the process. During *energy decay*, the atom must, as a requirement of returning to ground level, emit and dispose of the energy that was responsible for elevating the atom to its excitation level.

The emission and disposal of energy is accomplished in decay by emitting the energy to its surrounding environment of neighboring atoms. Thus, one atom might be excited by applied energy and dispose of this energy by dumping it to a neighboring atom or molecule. This process of transfer of energy is known as *reaction*. In this particular case, it is called *population reaction*.

Do not confuse this reaction with a *nuclear reaction*; the energy is not on the nuclear level, but on the atomic level.

Go back and reread this chapter before moving on. To understand the material in the following chapters, you must understand the terms defined in this chapter.

Chapter 3

Electromagnetic Wave Theory

TO GAIN A BETTER UNDERSTANDING OF THE LASER AND THE MASER, YOU need a working familiarity with the science of light and its propagation. Light is a specific type of electromagnetic wave, so this chapter covers the basic theories of electromagnetic phenomena.

FREQUENCY

Electromagnetic energy is propagated in the form of *waves.* All waves have some specific frequency. Waves, frequency, and the electromagnetic spectrum can be explained graphically, as shown in Fig. 3-1. In Fig. 3-1A, there is a drawing of a simple wave. In Fig. 3-1B, there are a number of complete waves with the distances between the *crests* (or peaks) of each wave marked.

It took just one full second to draw this wavy line. Because two complete waves (or *cycles*) were created in one second, the propagated wave has a frequency of two.

Frequency is a measurement of how often something happens over any given period of time. In the case of electromagnetic waves, the frequency is the number of complete cycles the wave goes through in exactly one second.

The wavy line in Fig. 3-1B consists of three peaks, forming two complete wave cycles, which alternate above and below the centerline (zero).

In the past, frequency was usually measured in *cycles per second* (cps). Although this term is still occasionally used, it is usually replaced by Hertz (Hz), as a lasting tribute to the scientist Heinrich Hertz. Hertz is known as the father of electromagnetic wave theory. Just remember that the term *Hertz* means exactly the same thing as cycles per second (Hz = cps).

As stated, the waves in Fig. 3-1 were drawn at a frequency of two complete cycles in a second, or 2 Hz. The distance between corresponding points on the waveform is referred to as *wavelength*. If you take a ruler and measure the distance between the peaks, you will find that the wavelength is 2 inches.

Moving the pen twice as fast as before results in four complete waves to be drawn in the same period (one second), as shown in Fig. 3-2. Now the wave has a frequency of 4 Hz. The wavelength this time is only 1 inch because twice as many cycles were drawn in the same total space as had been done in Fig. 3-1.

You can graphically represent almost any series of events. First you have to establish just exactly what is happening and which events you wish to record.

Electromagnetic waves travel at the speed of light. In a vacuum, this is a speed of 186,000 miles per second, or 300 million meters

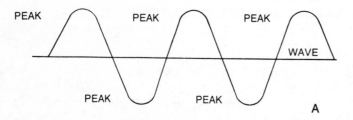

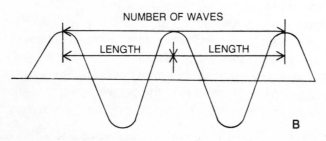

Fig. 3-1. An electromagnetic wave.

per second. These electromagnetic waves are composed of fluctuations similar to those shown in Fig. 3-1 and Fig. 3-2.

Each oscillation of a wave shows the energy that the wave carries. The magnitude of the rising and falling, or the value of energy, is very predictable. The frequency tells how many times the rise and fall of energy occurs in any given time period. Frequency also gives a direct mathematical indication of the wavelength.

This self-identifying phenomenon only requires the speed, number of cycles, and the *amplitude* (how far above and below the centerline the peaks are) before a picture can be drawn. Because the speed is constant (the speed of light), there are really only two variables to be concerned with.

An example of an electromagnetic wave with a frequency of 6 Hz is illustrated in Fig. 3-3. In one second, this wave has traveled 186,000 miles. To determine its wavelength, divide the number of cycles into the distance it travels in one second.

Thus:

$$\frac{186,000}{6} = 31,000 \text{ miles}$$

In discussing electromagnetic waves, the metric system is more commonly used. The same mathematics apply:

$$\frac{300,000,000}{6} = 50,000,000 \text{ meters}$$

Real electromagnetic wavelengths are not nearly as long as the example illustrates. Electromagnetic frequencies are much higher than in the example, and as the frequency increases, the wavelength decreases.

Actually, electromagnetic waves start at the lower end of their spectrum at about 10,000 Hz (10 kHz, or kilohertz). At this frequency, the wavelength is 18.6 miles. At the upper extreme of the spectrum,

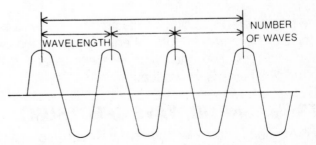

Fig. 3-2. An electromagnetic wave with twice the frequency of the wave shown in Fig. 3-1.

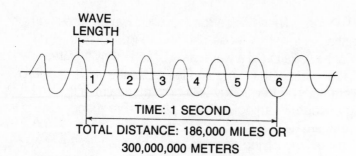

Fig. 3-3. An electromagnetic wave with a frequency of 6 Hz.

micro-microwaves exist. These waves are also known as *cosmic rays.* They occur at frequencies of millions of megahertz (MHz). (A megahertz is 1 million hertz.) These extremely high frequency waves have a proportionate wavelength of approximately 1×10^{-32} centimeter. Computations of wavelength and frequency can be made with this simple formula:

$$c = f\lambda$$

where c is a constant representing the speed of light (3×10^8 meters per second), f is the frequency (in Hertz), and λ is the Greek letter lamda, which denotes the wavelength (in meters).

Work through a typical example. What is the wavelength of the signal broadcast by a radio station transmitting on a frequency of 500 kHz (500,000 Hz)? Algebraically rearrange the equation given above, and substitute the known values:

$$c = f\lambda$$
$$\lambda = c/f$$
$$= 300,000,000 / 500,000$$
$$= 600 \text{ meters}$$

In the preceding discussion on electromagnetic waves, merely the footprints of this particular form of energy have been examined. A working investigation of the immediate effects and potential of electromagnetic waves is now necessary.

ELECTROMAGNETIC WAVE SPECTRUM

The illustrations of the electromagetic wave spectrum (Figs. 3-4 and 3-5) give a graphic illustration of the apparent continuity of the wave-frequency relationship. The spectrum runs from the longer-length

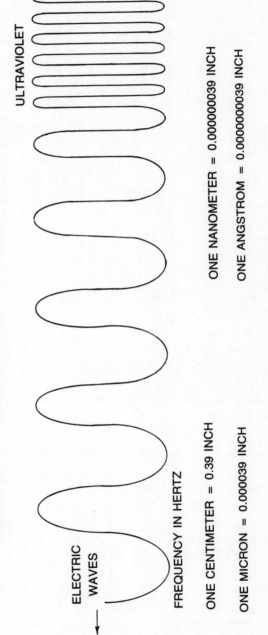

Fig. 3-4. The electromagnetic wave spectrum.

32

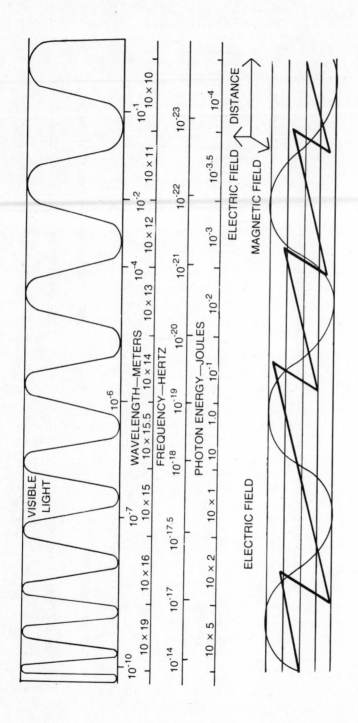

VISIBLE
LIGHT

WAVELENGTH—METERS

FREQUENCY—HERTZ

PHOTON ENERGY—JOULES

ELECTRIC FIELD / DISTANCE

MAGNETIC FIELD

ELECTRIC FIELD

Fig. 3-5. More on the electromagnetic spectrum.

radio waves at the far left to the higher-frequency, shorter waves at the far right. The interconvertible units of length are included as a reference because you will use them later in calculations. The wavelengths of most interest here are very short. You will deal with unusually minute lengths such as the angstrom, the micron and the nanometer. If you are not familiar with these units, refer to Chapter 15.

It is important to remember that all electromagnetic waves are similar in their properties. This means that any one wave is not absolutely peculiar to the specific device that generated it. Thus, an electromagnetic wave of 200 kHz propagated by the piezoelectric effect of a quartz crystal is identical to a 200 kHz electromagnetic wave generated by an electronic tank circuit (inductance-capacitance oscillator circuit). It does not matter by what manner or *mode* the wave is generated. An electromagnetic wave is an electromagnetic wave regardless of its origin.

Many devices or methods are capable of producing electromagnetic waves of some frequency. The energy level may differ, but characteristically its ''personality'' is created by any method that has the capability of propagation.

In all instances, the transmission and reception of electromagnetic waves are reversible and interconvertible between sender and receiver, and between receiver and sender. This means that an electromagnetic wave can be created electrically, transmitted through space, received in a resonant tuning circuit, transferred back through space to the original source of generation, and still maintain its exact physical characteristics in every detail.

Of course, the wave might lose some of its original energy level, which may have become exhausted in transit or lost through the reproduction of the wave in the transferring devices. These waves are simple, unencoded signals. If the electromagnetic wave is modulated with some sort of intelligence, data, or program, the carried information might be distorted in transit. There is also the possibility of random noise being added to the desired modulating program. However for a simple electromagnetic wave, such problems are not significant.

All electromagnetic waves are manifestations of energy. They were at one time created or propagated by a piezoelectrically vibrating crystal, a thermionically hot filament, the discharge of a capacitor, or some other method. An electromagnetic wave also can be propagated by an electron dropping from a higher to a lower level of photoelectronic energy or by the alternating field surrounding an electromagnetic field. The propagated wave would represent a percentage of the applied energy.

The applied energy is represented by the wave amplitude. Amplitude is the measurable vertical height or distance in either direction, perpendicular to the *time line* (x-axis) of the graph of the wave.

Up to this point, only the singular occurrence of one wave has been considered. Electromagnetic waves can, and quite often do, exist in multiples at any given instant. The waves can all be of the same frequency, or they can be different frequencies. If they exist as waves of different frequencies, they can be segregated by their respective frequencies. They can be individually tuned and dealt with independently of each other.

When two or more electromagnetic waves occur simultaneously and have the same frequency, they must be dealt with quantitatively. It must be determined if these waves are in phase or out of phase with each other. If they are out of phase, it must be determined by how much and expressed in degrees of cycle. One full cycle equals 360°, a half cycle equals 180°, a quarter cycle equals 90°, and so on.

The graphically expressed, simultaneous occurrence is called *not phase* and is a multiple cycle of waves. The graphic representation shows the difference in the *lag-lead* relationship between the two or more occurrences in respect to their starting points and stopping points, or in respect to points of interception along the time line of the graph.

To satisfy immediate and practical interests, turn your attention to waves of the same frequency. For example, waves of the same frequency include all of the waves of red light, which have a frequency of 4.3×10^{14} Hz. Waves of red light exist only at this frequency, and more often than not, they exist in groups of waves that are either totally or partially out of phase with each other.

AMPLIFICATION

When two or more waves occur simultaneously and exactly in phase with their group, there exists an *amplifying effect*. All of the values of the waves add up to the effective value of a much greater wave. Thus, many smaller values could be amplified, providing that all the smaller waves were in phase with each other. This phenomenon is exploited to a great extent in the amplification of light and microwaves, as you will learn in later chapters.

Examine the two waves that exist simultaneously and at the same frequency in Fig. 3-6. The only difference between them is that one wave has a greater amplitude than the other. But they are exactly in phase. You could say that the two waves are superimposed. In-phase

35

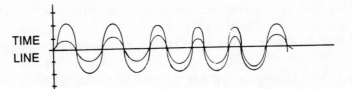

Fig. 3-6. In-phase electromagnetic waves that are additively superimposed.

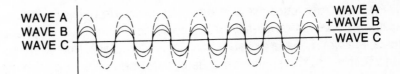

WAVE A
WAVE B
WAVE C

WAVE A
+WAVE B
WAVE C

Fig. 3-7. Wave C is the resultant wave of waves A and B.

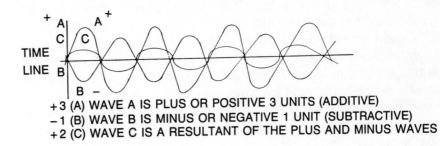

+3 (A) WAVE A IS PLUS OR POSITIVE 3 UNITS (ADDITIVE)
−1 (B) WAVE B IS MINUS OR NEGATIVE 1 UNIT (SUBTRACTIVE)
+2 (C) WAVE C IS A RESULTANT OF THE PLUS AND MINUS WAVES

Fig. 3-8. Electromagnetic waves in exact out-of-phase, subtractive superimposition.

waves are additively superimposed. Another example is illustrated in Fig. 3-7.

When two or more waves exist together, they are said to be superimposed. It is irrelevant whether or not they have the same frequency.

In the preceding examples and commentary, superimposed waves were considered. They were in the additive state, which would produce a resultant wave of a proportionately amplified value.

Superimposed waves also can be subtractive as shown in Fig. 3-8; waves A and B have exactly opposite amplitudes. The result is that the waves tend to subtract or cancel each other. If two waves of the same frequency are 180° out of phase with each other, they will cancel each other completely. The result is zero.

Bear in mind that if the waves are either exactly in phase or exactly out of phase, their resultant or effective amplitude is the algebraic collection of their respective plus or minus values. If the components are all plus, their resultant is an amplification. If the components are all negative, they will tend to cancel each other out. If the components are a combination of positive and negative values, they will produce an interference effect. This cancelling resultant will be less than the otherwise mathematical sum of the individual component values.

Only those waves that are either in phase or exactly out of phase have been evaluated so far in this discussion. Their extremities were either completely in phase, or completely out of phase.

As frequently happens, some of the waves are between the two extremes of being completely in phase or completely out of phase. That is, the waves have phase differences or angles somewhere between 0 and 180°. This type of situation can be handled by the use of phasors in drawing a resultant parallelogram. The resultant is found as the diagonal is formed by the phasors, or component length sides of the constructed phasor box. Plans for the construction of phasor parallelograms have been included in Chapter 8 of this book. However, it quite practically serves immediate and instructional purposes to limit considerations right now to the extent that waves exist either in phase or out of phase. Their individual or component values are merely added together. One wave with a value of one unit and another of two units equals three when added algebraically. All the numbers should be positive because they will be measured from the top of the time line.

The resultant sum of the waves is a product all its own, so to speak. It is calculated as though it were one resultant wave. A resultant wave exists as shown by its effects as a resultant quantity of *amplified energy* (Fig. 3-7).

If there are two waves exactly in phase, they exert an additive effect on the resultant wave of their creation. This additive action indicates that the waves or the amplitudes they represent have an amplifying effect on the resultant wave that previously and independently existed as a superposed component.

In any case, the resultant wave of two or more superposed component waves will yield—by the algebraic collection of their individual values (amplitudes)—a resultant effect either greater than or less than the sum of its components. This will depend on whether the components are additive (reinforcing) or subtractive (cancelling).

For intermediate phase relationships (phase angles between 0 and 180°), the effect at any instant might be either additive or subtractive. If the instantaneous value of both waves is on the same side of the zero line (either both positive or both negative), the result will be additive, at least at that instant. If instantaneous value of one wave is positive (above the zero line) and the value of the other is negative (below the zero line), the effect at that instant will be subtractive, with the larger value being dominant.

With intermediate phase relationships, the effects may shift back and forth between an additive and a subtractive resultant at various points during the cycles.

EVALUATION OF ELECTRO-MAGNETIC WAVE THEORIES

In this chapter, you studied the fundamental characteristics of electromagnetic waves. Waves were explained in respect to their graphic representation, their physical construction and nomenclature, and—most important—their proven nature to exist or manifest themselves in multiple quantities. Waves in multiple quantities exert either an additive or amplifying effect, or a subtractive cancelling effect on the resultant wave.

The terms amplify and amplification have been repeatedly used. These terms have been emphasized at appropriate times, where amplification was defined. It is by this process of amplification that lasers and masers are able to perform their extraordinary feats, and it is by the action of superposition of in phase energy waves that amplification occurs. Another level of progress into the study of the laser has been reached.

You have now a conversant knowledge of amplification, which is the characteristic property of electromagnetic waves. The journey to the heart of the system will continue by exploring the component branches of physics that produce the laser and maser. After examining enough of the components, you will consider the constructed product. But before then, there are several more concepts or components to deal with. The next one is the light wave theory, where much of your recently acquired knowledge of electromagnetic wave theory will be put to good use.

Chapter 4

Light Wave Theory

IN THE LAST CHAPTER, YOU WERE INTRODUCED TO THE ELECTROMAGNETIC wave spectrum. This spectrum is divided into several *regions,* or families. One such region is the family of light, including the submembers of infrared, visible, and ultraviolet light. In this chapter, you will study the family of light exclusively.

Ordinarily, when you think of light, you think of what you can see or sense, that is: *visible light.* Putting it more scientifically, visible light comprises those electromagnetic waves between the frequencies of 10^{13} to 10^{14} Hz and having respective wavelengths of 0.00004 to 0.00007 cm. Electromagnetic waves in this range give the impression of visible colors when received by the photosensitive retina of the eye. For example, the color red has a wavelength of 0.000066 centimeter, and the color violet has a wavelength of 0.000042 centimeter. Before proceeding further, study Table 4-1 for purpose of color orientation.

Every color is created by its own particular identifying wavelength and frequency. White light (like that given off by the sun or an ordinary incandescent light bulb) is a representation of all the possible colors at once (a combination of all the individual frequencies). White light is akin to white noise in the field of acoustics. The total effect of white light is that no particular color or frequency predominates. You see all the colors—the entire mixture of wavelengths—simultaneously.

Table 4-1. Wavelengths of Visible Light.

Infrared						Ultraviolet
nanometers (nm)						
1200	700	600	550	500	400	200
centimeters (cm)						
0.00012	0.00007	0.00006	0.00005	0.00004	0.00003	0.00002
Infrared	Red	Yellow	Green	Blue	Violet	Ultraviolet

Color	Wavelengths	
Red	0.000066 cm	660 nm
Orange	0.000061 cm	610 nm
Yellow	0.000058 cm	580 nm
Green	0.000054 cm	540 nm
Blue	0.000046 cm	460 nm
Violet	0.000042 cm	420 nm

Because while light is made up of all possible colors at once, it is sometimes called polychromatic light. *Polychromatic* means many colors. A pure, single-frequency light consisting of just a single color is called *monochromatic light.*

The family of light extends both below (ultraviolet) and above (infrared) the visible range. The family of light waves, including infrared, visible, and ultraviolet waves, have, in addition to the qualities of all electromagnetic waves, the following special characteristics:

• They are rectilinear in their propagated direction. *Rectilinear* means that the waves exist as rays of energy, behaving as if they were straight lines.

• They are able to undergo *refraction* or bend and have their direction diverted as they pass through a medium that offers optical resistance (a change in optical density). Waves refract in a very predictable manner and as a result of the difference in speed readjustment.

• They are capable of *interference,* the function of superimposing their presence on other nearby waves. Interference affects the wave itself and its neighboring wave(s).

- They are *reflective,* or able to be reflected from a boundary, or surface of a medium. Waves reflect with strict mathematical precision.
- They are convertible in that their energy can be converted or transferred from its original manifestation or existence as light to another form of energy (such as heat, either electronic or chemical). The existence also can be changed by the disturbance of molecules to an extent that could either impart their energy or convert it to another manifestation.

These functions and characteristics will be discussed in the following sections. But first, there is yet another facet of light to examine, coherence.

COHERENCE

Incoherent light is the light received from the Sun or from an incandescent light bulb. The light is radiated or sent out in all directions and consists of all manners of polychromatic waves. This all occurs simultaneously and interferes with itself. Because of this diversion, or spreading out, there is a loss of brightness and intensity as it progresses from its source to its final point of destination.

If all the complementary colors were eliminated from this gross and random bunch of multiple frequency rays or emissions, and just one primary color were left, a monochromatic, but still incoherent collection of light would remain. Totally polychromatic light is composed basically of the three primary colors of red, green, and blue. Thus, there could be a red, blue, or green incoherent light beam.

If this monochromatic beam of light were further reduced by reflecting it with a parabolic reflector into a parallel stream of rays, a monochromatic beam of light would result. It could penetrate its way into a medium farther than when it was just diverged into space. An example of this type of light is that which is produced by a search light.

Taking this monochromatic light and arranging all of the individual rays of its composite beam in such a way as to cause all the constituent and fundamental waves to vibrate, or oscillate in step with itself, will produce coherent light. The component waves are then constructively reinforcing each neighboring wave.

Coherent light is therefore, as far as the laser is concerned, monochromatic in phase emissions, or rays of radiant energy that are emitted in a beamlike manner. Because it is coherent, it is amplified in relation to the total of in-phase emissions. Because of this in-phase personality, the total effect is one that imparts a constructive character to the finished product. In this case, the constructive character would be the beam of emitted coherent light.

Now that the basic character, or identity, of light has been examined, you will see how it behaves under conditions of refraction and reflection. In a sense, you will see how light makes its way through what amounts to an optical obstacle course.

REFRACTION

The scientific definition of refraction requires a lengthy description. Fortunately, the discussion can be simplified quite a bit. Refraction is covered in sufficient depth to give you a practical, working knowledge of the principles involved.

In relatively simple terms, *refraction* is the property light waves exhibit when they pass through a medium or substance, such as a glass lens. Actually, the concern is with what happens when a light wave passes from one substance into a different substance. For example, a light beam might pass through air, and then enter a glass lens, or vice versa. Refraction is due to the difference between the optical density of the two substances. The light wave will bend, or refract, from its original, or incidental, line of travel to a new line of exit at a different angle. The exact angle of its *exit* compared to the angle of *entrance* into the second substance is a function of the refractive ability of the medium.

The refractive ability of any substance is defined by the *refractive index*. The index is the ratio of the speed of light in a substance to the speed of light in a vacuum. The refractive index of a vacuum is 1.0, or unity.

The exiting angle, or degree of refraction, is also directly dependent upon the *incident* angle, as drawn to a line normal on the surface of the lens. A *line normal* is a line that is perpendicular to a surface.

A typical example of refraction is illustrated in Fig. 4-1. As light waves meet the surface of the lens and enter it, the line of direction is bent upon its arrival into the refracting medium (the glass of the lens). As the light wave progresses through the lens, it breaks through the exiting surface and its line of direction is again bent to yet another angle. This refracting ability of any lens is caused by the retarding capacity of the lens material. This slows down the speed of light as it passes through the medium. It is the prompt and delibrate change in velocity that causes the break in the direction of travel of the wave. This turning, or bending, is not an external function, as in the case of reflection (discussed in the next section), where the phenomenon is more analogous to a ball bouncing off a surface. Refraction is a different concept in that it is achieved inside the medium, and not externally.

REFLECTION

Reflection involves a wave striking, or being incident to, a surface in a very predictable manner. The wave bounces, or reflects, off of the incidental or reflective surface. More specifically, the incident wave, or particle, will be turned back by a reflecting surface at the same angle of its arrival. The principle is exactly the same as a ball bouncing off a wall. Reflection is illustrated in Fig. 4-2.

For simplicity, Fig. 4-2 describes a nominal angle of 45° incident and an equal reflection. This angle, in respect to a line normal to its surface, could be any angle at all. Thus, if the incidental wave were to arrive at the reflecting surface at an angle of 36.8° (for example) to its line normal, then it would be reflected at 36.8° from its line normal.

In the simplest case, reflection uses a flat surface, like an ordinary mirror. In laser applications, spherical and parabolic reflective surfaces are often used.

The function and behavior of a light wave will, regardless of the geometry of the reflective surface, obey all the laws and mathematical rules that apply for them in flat, or straight-line situations. Reflection from a curved surface is illustrated in Fig. 4-3.

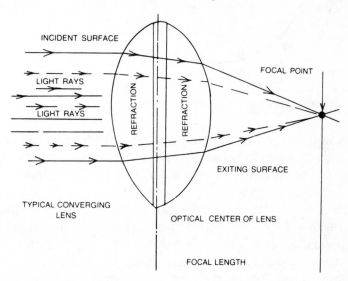

Fig. 4-1. Refraction angles of incident rays entering and exiting from a converging lens.

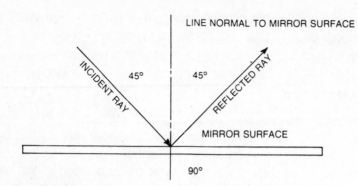

Fig. 4-2. An angle of incidence and reflection.

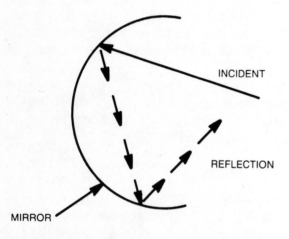

Fig. 4-3. Reflection from a curved surface.

Now consider the causes of the actions of refraction and reflection. One such cause is the fashion in which light waves travel either through a medium or vacuum in a *rectilinear*, or straight manner.

Rectilinear propagation is one of the foremost problems encountered by founding scientists in their exhaustive quests for explanations of light-wave phenomena. Rectilinear propagation is the ability and manner in which light can manifest itself by existing as both transverse waves and as individual particles of energy.

With the knowledge that light rays do exist as particles, you can more easily understand further theories on the functions light rays can perform.

Light rays are rectilinear in their travel or propagation. Consequently, they form straight lines, and simple geometry can be applied in calculating their reactions to, and interactions with, any surface or medium that would interrupt their intended or incidental path of travel. Examples of such media or surfaces include the surface of a mirror or the internal and external boundary surfaces of a refracting lens or some other transparent medium.

Because of the rectilineal straightness of rays and the inherently geometric advantages of description, mechanisms for their use can be designed with mathematical precision. It is possible to predict the reactive angles from a straight-line course because light rays are rectilineal, almost as if they were exact lines that could be drawn with a pen or pencil. This phenomenon can be used to great advantage in designing mirrors and lenses and their functions.

INTERFERENCE

The character of *interference* of light will be considered, and it will be evaluated only as the resultant effect on wave intensity in respect to the energy level. The wave is affected while in transit from its point of emission or propagation as well as at its point of destination.

Due to the existence of multiple simultaneous waves, interference is the action by which one or more waves will tend to enhance or cancel the effective amplitude of another wave. This is the same principle as is involved in combining in-phase and out-of-phase electromagnetic waves. Interference occurs in all instances where multiple waves coexist. The only difference between additive and destructive interference is in the manner in which they exert their influences. The effect of *constructive interference* is the amplification of the emission as a whole. The effect of *destructive interference* is the reduction of the emission as a whole. Interference exists constantly when two or more waves get together.

Interference also has an influence on the chromatic effect of electromagnetic waves. In addition to the fact that light rays are being treated as particles, they are first and always, energy—that is, electromagnetic waves.

Once you understand the concepts of reflection, refraction, rectilinear propagation, and interference, you can describe the laser beam. The laser beam is an emission of *collimated* (parallel, pencillike, nondiverging rays) and coherent (all waves in phase with each other) light energy. The light energy may be in the infrared, visible, or ultraviolet regions of the electromagnetic wave spectrum.

Chapter 5

Mirrors and Lenses

AS YOU HAVE SEEN IN THE PRECEDING CHAPTERS, LIGHT RAYS (IN FACT, ALL electromagnetic waves) can be made to perform a number of tricks. Certain characteristic functions can be utilized to create certain phenomena through the use of mirrors and lenses.

The characteristics of light include reflection and refraction. These characteristics can be proven and exploited by the use of mirrors and lenses, respectively.

MIRRORS

A mirror is a highly reflective surface. There are three basic types of mirrors:

- flat mirror
- spherical mirror
- parabolic mirror

The *flat mirror* (Fig. 5-1) is undoubtedly the most widely known type of mirror. It is the same kind you look in every morning. This type of mirror consists of a flat piece of glass that is silvered on the back side. The front is the reflective side. By reflecting light, the mirror

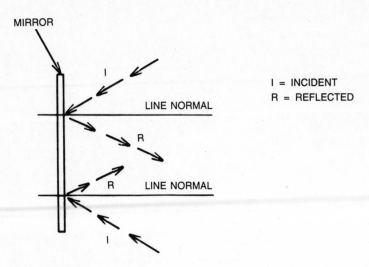

Fig. 5-1. With a flat mirror, the line normal lies at a right angle to the plane.

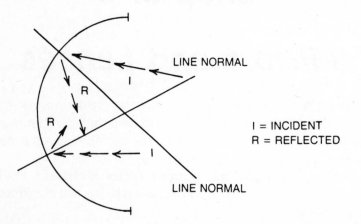

Fig. 5-2. With a spherical mirror, the line normal is the geometric radius of the mirror.

produces an image that directly duplicates whatever is in front of it.

Notice that in Fig. 5-1 the line normal to the flat mirror lies at exactly right angles (90°) to the plane of the mirror. This is true of all flat mirrors. As stated in Chapter 4, the angles of the incidental (incoming) and reflected (outgoing) rays of light are measured from this line normal.

A *spherical mirror,* as shown in Fig. 5-2, is half a sphere that has been silvered on the outer surface. A spherical mirror is reflective on the inside (concave side).

The third basic type of mirror is the *parabolic mirror,* which is illustrated in Fig. 5-3. This type of mirror is used almost exclusively

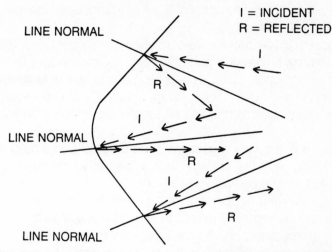

Fig. 5-3. With a parabolic mirror, the line normal is the line of parabola.

in searchlight reflectors and microwave transmission antennae. Although this type of mirror is not normally used in the construction of lasers or masers, it is mentioned here to complete the list of mirror types.

Because spherical mirrors are the most widely used type of reflector in laser applications, a more thorough discussion of this device is needed.

THE SPHERICAL MIRROR

The most useful type of mirror in laser applications is the spherical mirror. The line normal for this mirror is at all times the actual radius of the sphere (Figs. 5-4 and 5-5). As with all mirrors, all arriving incident rays falling upon its reflective surface will be reflected at an angle equal to the incident angle of arrival. These angles are measured from the line normal—the radius of curvature.

Because of the concave curvature, the reflected rays have the ability to meet, or *focalize,* themselves at a certain distance from the mirror surface. The point at which the waves meet is called the *focal point.*

If certain parameters of curvature geometry are followed and if the light source is a precise distance from the mirror, all the light rays would be returned—or reflected back—to the point of creation, called the *image.*

In Fig. 5-5, a spherical mirror is shown as a cutaway. This diagram also illustrates the radius of curvature, which is also the line normal for the mirror. Also shown are the *focal length,* or the distance at which the rays will meet, and the *principal,* or the point at which the radius of curvature is centered.

Notice in Fig. 5-4 that the random angles of incident rays, although always obeying the laws of reflection, are not meeting, or focalizing, at any one particular point. This is due to the lack of singleness of projection. All of the rays could be focalized to a definite focal point were such light sources maintained for the issuance of the light rays (Fig. 5-5).

In Fig. 5-6, the light rays have been created and aimed toward the spherically reflective surface. In this case, the reflected rays tend to meet at a single point.

Using two mirrors (as illustrated in Figs. 5-7 and 5-8), the point of image and curvature of reflection could be adjusted to reflect the rays of one mirror to become the image for the second mirror. The second mirror would then return the rays to the first. This receive-and-return process would continue as long as there were enough light from the source (Fig. 5-9).

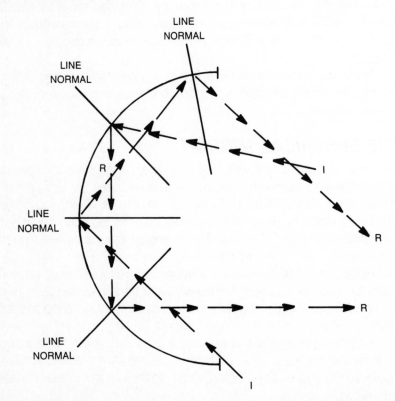

Fig. 5-4. A spherical mirror illustrating the geometry of incident light rays and reflected light rays to the line normal of the mirror. The line of reflection is always equal to the line of incidence.

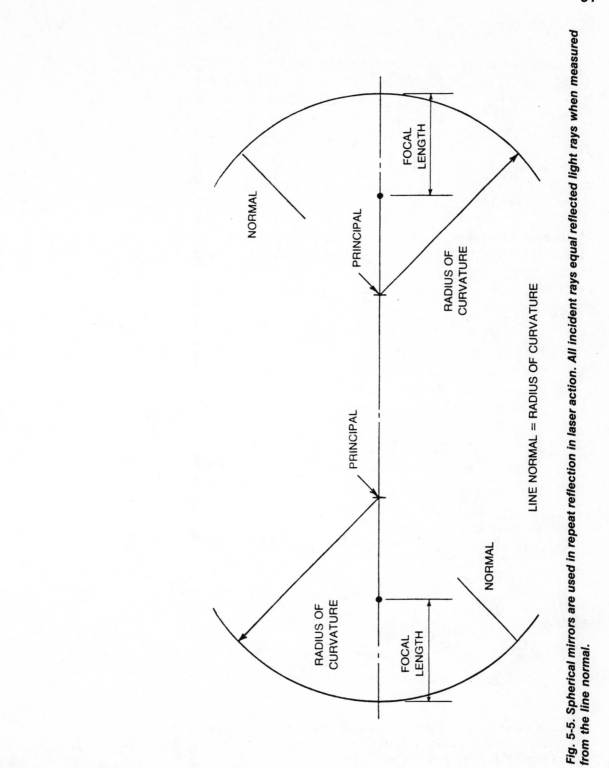

Fig. 5-5. Spherical mirrors are used in repeat reflection in laser action. All incident rays equal reflected light rays when measured from the line normal.

52

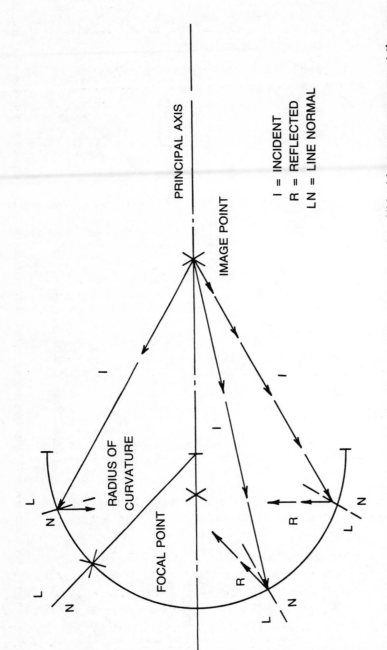

Fig. 5-6. A spherical mirror illustrating light rays being created at a fixed image point. All incident rays converge at the same focal point.

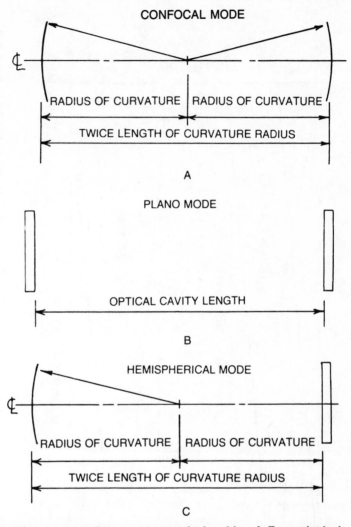

CONFOCAL MODE

RADIUS OF CURVATURE | RADIUS OF CURVATURE

TWICE LENGTH OF CURVATURE RADIUS

A

PLANO MODE

OPTICAL CAVITY LENGTH

B

HEMISPHERICAL MODE

RADIUS OF CURVATURE | RADIUS OF CURVATURE

TWICE LENGTH OF CURVATURE RADIUS

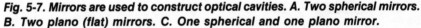

C

Fig. 5-7. Mirrors are used to construct optical cavities. A. Two spherical mirrors. B. Two plano (flat) mirrors. C. One spherical and one plano mirror.

An internal action is caused by the repetitively reflected rays that occur when a lasing or masing action is in progress, as shown in Fig. 5-10. The rays are reflected between the two mirrors, A and B. These reflectors form the cavity-end walls of the laser or maser. Mirror B is deliberately made a little less reflective than A. This provides a window for the exit of the produced laser beam. In the first gas lasers, the mirrors used were placed outside the gas tube at each end. Due to advances in glass blowing, the two end mirrors can now be sealed within the gas-filled tube. This is done after the mirrors have been prop-

54

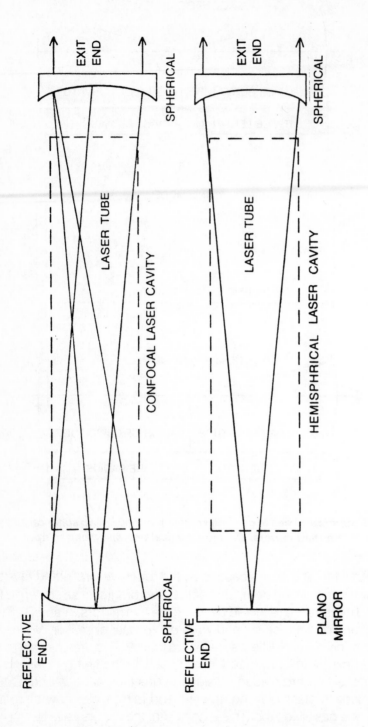

REFLECTIVE
END

EXIT
END

SPHERICAL

LASER TUBE

CONFOCAL LASER CAVITY

SPHERICAL

REFLECTIVE
END

EXIT
END

SPHERICAL

LASER TUBE

HEMISPHRICAL LASER CAVITY

PLANO
MIRROR

Fig. 5-8. Laser cavity showing mirror arrangements.

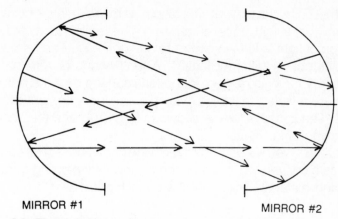

MIRROR #1   MIRROR #2

Fig. 5-9. The repetitive reflection of two mirrors.

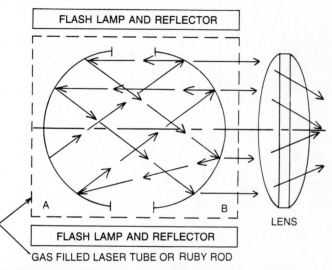

FLASH LAMP AND REFLECTOR

A   B

FLASH LAMP AND REFLECTOR

GAS FILLED LASER TUBE OR RUBY ROD

LENS

Fig. 5-10. Repetitive reflection as employed in a laser cavity.

erly designed and spaced at an appropriate distance from each other. This trend allows the convenient assembly and experimentation with gas lasers. This is a tremendous benefit because in previous designs, the slightest bump would result in just enough misalignment to render the laser nonfunctional. Needless to say, this caused a lot of headaches for experimenters.

LENSES

Just as a mirror is an optical device that exploits the properties of reflection, a lens is a device that exploits the properties of refraction.

Any lens, no matter of what substance it is made, refracts because it slows down the light rays entering it. This ability to retard the transmission of light is known as the *refraction index.* The refraction index is different for different lenses. A lens bends or refracts the direction of ray travel in respect to the lens surface and line normal.

There are a variety of lenses available, but there are two distinct categories. The *concave* lens is shown in Fig. 5-11, and the *convex* lens is shown in Fig. 5-12.

In laser work, you will deal mostly with the convex lens. The convex lens imparts a converging effect on the exiting rays passing through the refracting material. This is illustrated in Fig. 5-13.

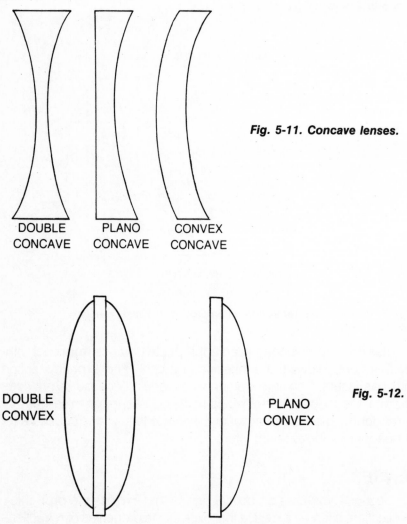

DOUBLE   PLANO   CONVEX
CONCAVE   CONCAVE   CONCAVE

Fig. 5-11. Concave lenses.

DOUBLE
CONVEX

PLANO
CONVEX

Fig. 5-12. Convex lenses.

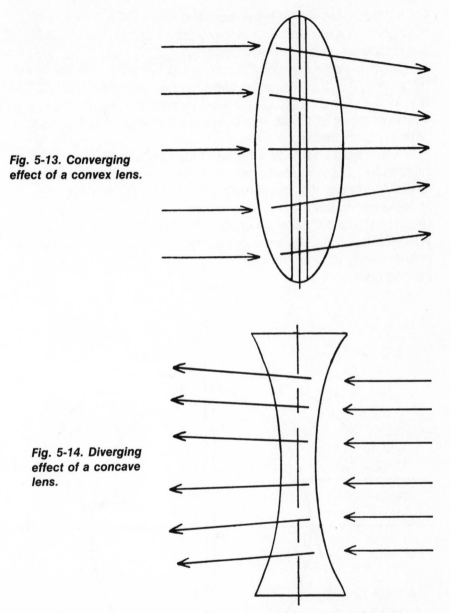

Fig. 5-13. Converging effect of a convex lens.

Fig. 5-14. Diverging effect of a concave lens.

The concave lens, on the other hand, causes the exiting rays to diverge, as shown in Fig. 5-14. The concave lens has no real value in the type of experiments and systems discussed in this book.

THE CONVEX LENS

There are several styles of convex lenses available. Each has its own special characteristics and properties for which it was ground.

For the purposes of this book, concentration will be on the double convex converging lens. This lens probably has the best features for this discussion of any lens obtainable.

A *double convex* lens is shown in Fig. 5-15. This lens is made by grinding a suitably refractive piece of glass to a prescribed radius of curvature. This radius of curvature is similar to the radius of the spherical mirror in that it has a radius extending from the point of principal at its center.

A lens must be selected that will refract the rays to a near-perfect point to focalize the greater portion of light rays. This characteristic is vital for a laser that is to create a heated focal point hot enough to melt steel. In that case, a lens is needed that can refract the laser beam to a focal point about 4 to 10 inches from the center of the lens. This would concentrate the collected beam and aim it at a point far enough away to keep the lens from melting or deforming from the intense heat.

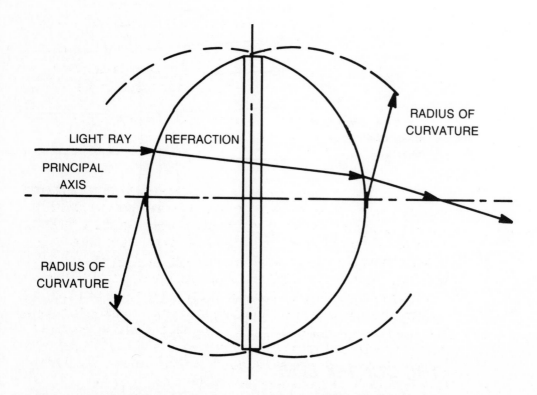

Fig. 5-15. A double-convex lens construction showing the cause of convergence.

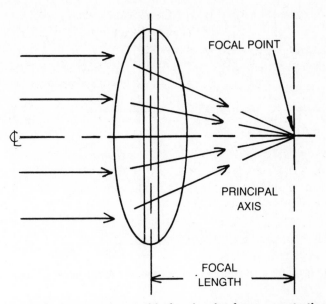

FOCAL POINT

PRINCIPAL
AXIS

FOCAL
LENGTH

Fig. 5-16. A single lens produces this focal point for concentrating rays.

The questions that must be answered for the projects in this book include:

- Will the lens concentrate the light as with a convex-type lens, or will it diverge the light as with a concave-type lens?
- What will the lens focal point be?
- If the lens is one of two or more lenses used in teleprojection, will it be optically compatible with the component lenses?

Obtain a good book on lenses before undertaking any extensive projects involving special arrangements of lenses. Most suppliers of scientific apparatus will be more than glad to provide you a catalog of the optical manufacturer who supplies their stock. Using the catalog makes filling orders easier because every manufacturer has different specifications and coded data for their lenses and optical supplies.

A list of suppliers and their addresses is given in Appendix E.

Where *compounding* (use of multiple lenses) is necessary, make a simple tube with suitable lens retaining rings. Find a tube with an inside diameter that will accommodate the lens outer diameter. To position the lens, inset plastic snap rings in front of and behind the lens. Secure the rings to the inside of the tube.

The tube that houses the compound component of the lens group will have an inside diameter at the first lens end that will provide a

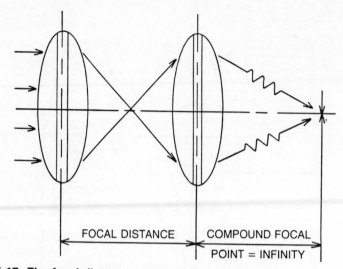

FOCAL DISTANCE | COMPOUND FOCAL
POINT = INFINITY

Fig. 5-17. The focal distance and compound focal point of two lenses.

slip fit over the laser tube. This arrangement allows you to move the tube longitudinally to adjust the focal length for beam projection. It makes the exit beam adjustable from twice the lens focal length (as shown in Fig. 5-16) to infinity (as shown in Fig. 5-17). Infinity means just that. Lasers have even shot their beams to the Moon and back with only negligible divergence.

Chapter 6

Laser and
Maser Concepts

LIGHT RAYS CAN PIERCE DIAMONDS—VAPORIZING THE TOUGHEST MATE-
rial known—and they can be gentle enough to penetrate the human
eye and fuse a torn retina to prevent blindness. Invisible microwaves
can forewarn us of approaching enemy missiles, and might soon replace the seeing-eye dog as an electronic vision system. These are
only a few of the present and expected feats to be performed by the
laser.

The maser, which was discovered before the laser, can be thought
of as a nonoptical laser. It is an electronic device very similar to the
laser, but it is different in that the maser emits microwaves rather than
light. The only real difference between them is the portion of the
electromagnetic spectrum that each uses.

Both the laser and the maser are electro-physical systems capable
of receiving, or being acted upon by very weak signals. This process
is known as lasing (or sometimes masing), and it results in a final output
reaction to the initial stimulation. Lasing brings about a desirable and
usable emission of an electromagnetic wave. The wave is coherent
and amplified and is capable of sophisticated tasks.

ABSORPTION AND EMISSION OF RADIATION

To understand how lasers and masers absorb and emit radiation, you need to understand that atoms and molecules are not really inert units. Think of them as bundles of energy. The electrons are very minute particles of electricity that are positioned around the nucleus by certain proven forces. These forces are established by the nucleus of the element and can be externally affected. Lasing (and masing) action depends upon this principle (Figs. 6-1, 6-2, and 6-3).

The atoms and molecules in a solid vibrate in a fixed position by virtue of their physical lattice. The lattice is an architectural aspect of the crystal. The atoms and molecules of liquids and gases are more movable; they can leave their positions and wander without restraint within the containing boundaries of their container.

Although the following description is not strictly accurate, it should help you visualize the concept. In a solid, the atoms and molecules are tightly packed together. There is not much room for them to move about, and they tend to stick to one another. They do move but not very much. Atoms and molecules in liquids and gases, on the other hand, have a lot more elbow room. There is more space between adjacent atoms or molecules, and they do not cling to one another quite so tightly. A liquid is midway between a solid and a gas in terms of atomic freedom.

All atoms and molecules, of just about every element and compound investigated to date, have one prominent characteristic. They are affected by light energy, magnetic forces, electrostatic or current charges, thermal influences, or in some instances mechanical reactions. This inherent personality of matter—to be internally affected by external effects—is utilized in both lasers and masers. It is called tuning. Actually, tuning is just another term for affecting something. It is as if you were to tune your television to a channel. The end result is that you have affected or influenced the electronic resonance of the circuit filtering aspects. Because you did so externally to the set itself, it is referred to appropriately enough as tuning.

Atoms and molecules can exist in different energy levels, separated from one another by definite steps. Suppose that radiation from outside acts upon a substance. If a given atom of the substance is at a low energy level, it will absorb the energy. The incoming radiation will be decreased by that amount. If the atom is at a high energy level when it is stimulated by radiation of the appropriate frequency, it will drop to a lower energy state. It may even do so without any specific stimulation by a wave—that is, spontaneously. Ultimately, it will emit radiation corresponding to the stimulated wave. Or, in the absence

63

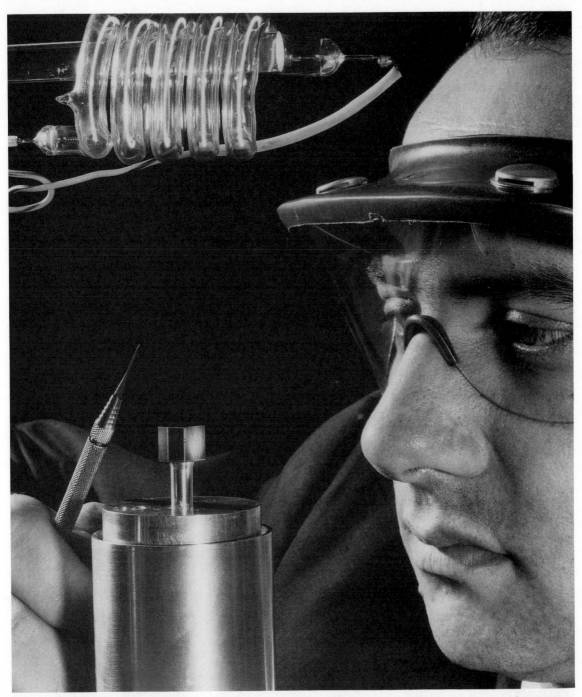

Fig. 6-1. A scientist views a cube of synthetic ruby crystal that forms the heart of a laser and generates the light. The light source (top) is used to excite the tightly packed atoms in the ruby, which amplify the light into an intense parallel beam.
Photograph courtesy of Hughes Aircraft Co.

64

Fig. 6-2. A synthetic ruby crystal (top) glows with absorbed light. The light source (below) pours random waves of light into the ruby.
Photograph courtesy of Hughes Aircraft Co.

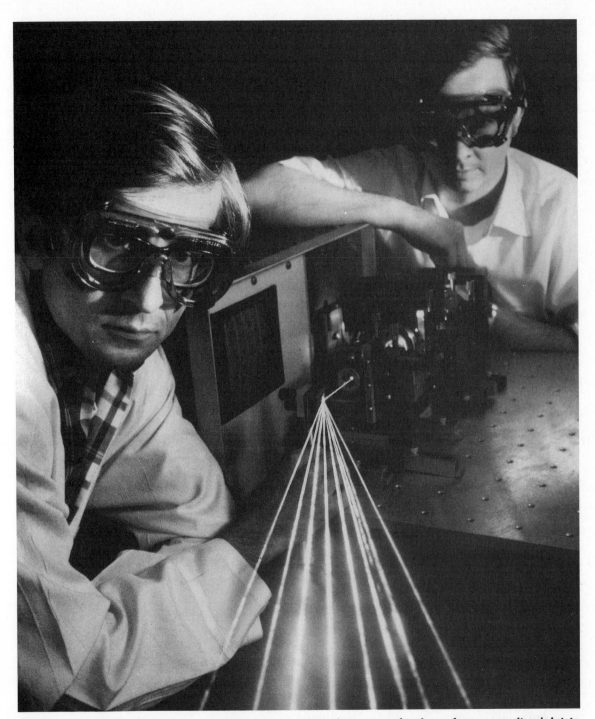

Fig. 6-3. Two Bell Laboratories scientists demonstrate the range of colors—from near-ultraviolet to yellow—that can be created from a dye laser called an exciplex laser.
Photograph courtesy of Los Alamos Scientific Laboratory.

of a stimulated wave, the emitted radiation will correspond in frequency to the difference between the higher and lower energy levels. Where there is stimulating radiation, this will be reinforced by the additional radiation. Under ordinary conditions, most materials will absorb much of the energy falling upon them. This is because most of their atoms are in the lower energy level to begin with.

The absorption and emission of radiation is normally a disorderly and random process. For one thing, the incoming radiation is usually a mixture of frequencies. Moreover, as atoms and molecules vibrate, they interact with one another, making the energy transformations even more complex. What if the atoms could be made to behave in a more disciplined way? What if they would jump up and down the energy steps at the physicist's command in unison like a troop of soldiers? That is precisely what has been accomplished with the laser and maser.

BASIS OF LASING AND MASING ACTION

The medium used in a maser can be a gas, a solid, or a liquid. The liquid medium has been selected for engineering reasons because its atomic or molecular structure has favorably arranged energy levels. The medium is led into, or is enclosed in, a cavity resonator. This is simply a container from whose walls incoming radiation can be reflected effectively. It also can be detained for a short enough time to affect the required properties given to the energy before its release as a beam.

First, this contained lasing or masing medium will have its population of atoms or molecules raised in average level of energy. This is in contrast to the level normally existing. This inversion can be brought about in several ways.

In one method of inversion, the low-energy atoms are filtered out through the use of an appropriate electrical field. This was the method used in the first operational maser, the ammonia maser, devised by J.P. Gordon, H.J. Zeiger, and C.H. Townes.

In another method, the temperature is substantially lowered with liquid helium or nitrogen, thus bringing the bulk (or population) of the atoms and molecules down to a low-energy state. Then these atoms are raised at the same time to the high-energy state by pumping. *Pumping* is accomplished by introducing into the system electromagnetic energy that is at a different wavelength than the stimulating wavelength that will be used later.

The next step is to introduce a stimulating wave at one wall of the cavity resonator. The stimulating, wave is a microwave of the appropriate frequency. The wave will then excite a number of the high-energy atoms in its path. Each of them will consequently emit electromagnetic energy of the same wavelength as the stimulating wave. Thus, the energy of the stimulated atoms will be added to the energy of the original microwave. The wave will then go into resonance. It will proceed to bounce off of one cavity-resonator wall and then the other in a process that progressively excites more atoms. As the wave continues to bounce back and forth from one end of the cavity to the other, the wave becomes progressively stronger. By the time the wave is emitted from the maser cavity, or when it breaks through the less-reflective wall or mirror, it will have become enormously amplified. This amplified microwave will reproduce unbelievably weak signals (electromagnetic waves) with great fidelity and very little noise or interference. In the ammonia maser, only two energy states are involved. However, in most modern masers and lasers, the energy reduction steps are based on three or more energy levels.

There are fundamental similarities as well as differences in the operation of masers and lasers. In the laser, atoms are raised to a high energy level state before they begin to emit. Unlike the maser, the laser emits a beam of intense light. It is infrared, or ultraviolet, radiation instead of microwaves that shows the different working regions of the electromagnetic wave spectrum. The laser does not usually function as an amplifier of a weak signal that might be impressed on it from the outside. The laser can be adapted as an amplifier of light signals coming from outside the proper frequency, by pumping it to a point where it would begin to lase by itself.

Light or electrical energy is used to raise the energy level of the atoms in the laser material. The now-activated, high-energy atoms begin to drop back to low-energy levels, which is called decay or de-excitation. As they decay, the atoms begin to give off the characteristic laser radiation or beam. Because this beam is rather weak at first, it must be reflected back and forth repeatedly between the end mirrors. (One mirror is at each end of the resonator cavity, as in the maser.) During this resonance, the beam stimulates other high-energy atoms to drop to their lower energy states, causing them to emit their contribution of energy of the same quality. This reinforces, or amplifies, the beam internally. When the beam becomes sufficiently energetic (in a tiny fraction of a second), it passes out of the laser through one of the end mirrors. This mirror is somewhat more transparent than the other end mirror, and is made less reflective. Think of this mirror as the window end of the resonator.

The beam of light emitted by the laser is almost absolutely monochromatic; it has a very narrow frequency range. Additionally, the beam is collimated and coherent, in reference to its nondiverging characteristic and its narrow frequency range. It has no other component frequencies to interfere with itself or to reduce its brilliance or power.

By subjecting lasing materials to various types of electrical, magnetic, and even sonic energy, laser output can be modified in various ways for the purpose of signalling, communicating, and experimenting. Many different patterns can be imposed on a beam, and even its color can be tuned or changed by an external tuning, such as in tunable dye lasers. Bear in mind that the only real difference between maser and its child, the laser, is its working region of the electromagnetic spectrum. The maser works in the microwave or nonvisible regions. The laser works in the higher frequency regions of infrared, visible, and ultraviolet. The maser is a nonoptical laser, and conversely, the laser is an optical maser.

LASER FUSION

There are several science laboratories that work extensively with lasers. One such place is the Lawrence Livermore Laboratory, which is operated for the Energy Research and Development Administration by the University of California.

Scientists at Livermore conducted experiments with a 2-trillion W Argus laser, one of the world's most powerful lasers and the third of a four-generation series of fusion research lasers at Livermore.

A master oscillator generates a 10 MW laser pulse lasting less than 1 billionth of a second. The pulse moves and is split in two. The pulses move in opposite directions. At the corners of the laser, they are reflected along parallel paths toward the target chamber.

During their journey from the master oscillator to the target chamber, the pulses pass through a series of amplifiers, filters, and devices that protect the laser equipment from being damaged by the extremely intense light. In addition, the pulse is diagnosed several times along the way for later analysis.

The Argus facility enables Livermore researchers to make studies of fusion microtargets compressed to high density. Such studies are an important step on the way to the goal of demonstrating the scientific feasibility of laser fusion.

A 1,000-joule, neodymium glass laser chain for controlled thermonuclear fusion research is also under development at the same laboratory. Increasingly powerful laser experiments conducted through the 1970s, culminating in 10,000-joule experiments scheduled to test the feasibility of fundamental/laser-induced fusion concepts. These experiments include the generation of micro-thermonuclear explosions in tiny heavy hydrogen pellets irradiated from all sides by extraordinarily precise powerful pulses of laser light.

Chapter 7

Introduction to the Laser

NOW, AT LAST, YOU CAN STUDY THE SPECIFICS OF THE LASER. THE ENTIRE system is nothing but a synthesis of the individual parts or branches of physics already studied in the previous chapters of this book. The first part of this chapter is something of a review, but one in which you will be able to construct—conceptually—a complete laser mechanism from the ground up.

CONCEPTUAL CONSTRUCTION

In Chapter 4, coherent and incoherent sources of light are discussed. As was mentioned, the atoms that make up the light source emit this radiation in the form of photons. In an incoherent source, the photons are emitted in a manner called *spontaneous emission.* In such an emission, the photons are emitted randomly in time and direction and have a wide range of wavelengths (colors). Thus, the light wave pattern of an incoherent source is analogous to the completely irregular wavelets produced by drops of rain on the surface of water.

On the other hand, the wave pattern of a coherent source, such as a laser, is like that of regularly spaced, parallel water waves being driven by a uniform wind. This regularity derives from the process of stimulated emission that occurs in a typical laser. Such an emission arises from the action of electrical or light energy on any of a variety of gases, solids, or liquids. The individual material used is contained in a suitable confining enclosure called a resonator.

When an electromagnetic wave, or light, of a particular wavelength passes through a suitable, highly excited (energized) material, it stimulates the emission of more electromagnetic waves of the same wavelengths. This light, or radiation, is in phase with, and in the same direction, as the stimulating light source. The resulting high-intensity (amplified) and coherent beam is the end product of laser action. Through this medium, some astounding scientific experiments can be performed.

Absorption, Spontaneous Emission, and Stimulated Emission

The three basic interactions of photons with matter are illustrated in Fig. 7-1. Atoms or atomic aggregates can contain only certain discrete amounts of internal energy. They can exist only in certain discrete energy levels.

An atom will normally reside in the lowest energy state possible (called the ground level) unless it is given energy by some external means, or is stimulated. One way an atom can gain energy, or become excited, is by the absorption of a photon of light (Fig. 7-1A). However, such absorption by an atom can take place only when the incoming photon has an energy exactly equal to the energy level separation E in that atom. Referring to Fig. 7-1, for instance, $E = E_1 - E_2$. If left to itself, an atom can lose energy by spontaneous emission, radiating a photon of energy E in any direction.

An excited atom can also be stimulated to emit a photon of energy E, if another photon of energy E strikes that atom. As a result, two photons will leave the atom: the original photon striking the atom and the new photon emitted by the atom. Most important, the two photons will have the same wavelength, the same phase, and the same direction. Thus, the stimulated emission process (Fig. 7-1C) is the basis of laser operation. It functions as a coherent amplifier on an atomic scale.

Population Inversion

For a laser to function, stimulated emission must predominate over absorption throughout the laser medium. The probability per unit time of the occurrence of each of these processes is the same. This was theoretically proven by Einstein in 1917. Therefore, for stimulated

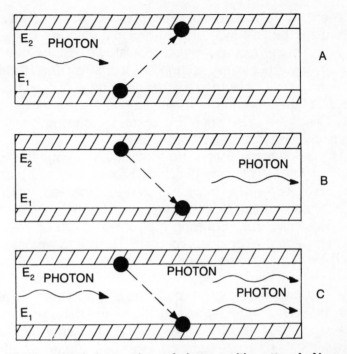

Fig. 7-1. The three basic interactions of photons with matter: A. Absorption. B. Spontaneous emission. C. Stimulated emission.

emission to predominate, more atoms must be forced into the excited (high-energy) state than are left in the lower-energy state. The distribution is called population inversion, or the reverse of the normal ratio of excited to nonexcited atoms in the mass population.

Mass Excitation

An inverted population can be produced or achieved by energy pumping. In insulating-crystal lasers, this pumping is accomplished by the application of intense radiation with light of a higher frequency than what is desired to amplify in the laser. This process is called *optical pumping,* or mass excitation.

In semiconductor junction lasers, the pumping is brought about by the use of electronic currents. It is thus called *electronic pumping.* In gas lasers, the pumping is achieved by electron-atom, or atom-atom collisions, and is thus referred to as *collision pumping.* In chemical lasers, the excited atoms or molecules of the medium are excited by means of chemical reactions. This is called *chemical pumping.* In some gas-dynamic lasers, the pumping is accomplished by means of the supersonic expansion of a gas, which is called *gas expansion pumping.*

There are several possible means by which the population of a medium can be inverted. Each means is dependent upon its own particular system. Once an inverted population exists, the laser is ready to amplify any light passing through it, or that which it can be triggered by. To use such a medium as a laser light source, or optical oscillator, the optical gain arising from the inverted population must be in excess of the loss in the resonator. (The losses are caused by scattering, diffraction, and reflector imperfections.)

To provide amplification, the optical path through the laser material is made long in one direction. Highly reflecting mirrors are arranged at each end so that they can repeatedly send or reflect the light back and forth many times along the laser length before the light is released from this resonating medium or enclosure cavity. The mirrors provide the necessary feedback, similar to what is required in rf (radio frequency) oscillators. The feedback mirrors convert the optical amplifier into an optical oscillator. Oscillation takes place only in the enclosed medium and is bounded by the mirrors. To allow the eventual release of the trapped light, one end mirror is deliberately made slightly less reflective. Alternatively, one mirror can be made somewhat transparent to allow the laser beam to emerge from the resonator once it has been sufficiently amplified.

One of the first practical laser prototypes was the *ruby laser*. This unit is shown in a simplified diagram in Fig. 7-2. The basic components of a pulsed ruby laser include:

- A light-emitting medium, or substance
- A ruby crystal
- Two reflective mirrors that face each other and provide the confining barriers for the medium
- The resonating cavity
- A controllable source of energy to excite the medium—in this case, a xenon-filled flash lamp (in a spiral form about the diameter of the ruby crystal)

The pulsed ruby laser operates as follows:

- An electrical pulse or charge is set through the flash lamp, causing it to emit a burst of white light.
- A portion of this light is absorbed by the chromium atoms in the crystal.
- The atoms, in turn, emit a red light in all directions.
- That portion of the red light that strikes the end mirrors is reflected back and forth many times.

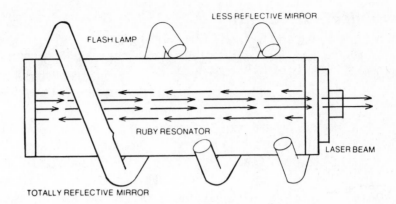

Fig. 7-2. An elementary ruby laser and its components.

- During reflection, the red light is being amplified.
- Some of this red light passes completely through the one slightly less mirrored end of the resonator.
- It comes out as a highly monochromatic, directional, and coherent beam of collimated light energy.

The preceding description is very elementary and nontechnical. More detail is provided in subsequent chapters.

LASER CHARACTERISTICS

The shape of the resonator and the shortness of the wavelength of light make the laser emission look like a pencillike beam. A narrow beam is produced because many of the excited atoms are stimulated to emit in a specific direction, rather than in random directions. This accounts for the extreme brightness of laser light. All of the light energy is concentrated into a narrow beam, rather than being spread out.

Laser light is also very nearly monochromatic. Its frequency spread is sometimes as small as a mere one-trillionth of its generated frequency.

The use and functions of lasers depend on these characteristics. The intense brightness, or concentrated power, is the most important property for cutting, welding, and drilling. This is especially true when working with materials that would be difficult or impossible to work with using traditional machining techniques. The monochromatic quality of a laser is most important in applications such as spectroscopy. The shortness of the pulse is of paramount concern for high-speed uses of the laser in special fields of photography. The combination of direction, high power, and short pulses make possible such long-distance measurements as lunar ranging.

Amplification Control

Instantaneous and high-power outputs can be gained from a laser by using a technique called *Q-switching* of the resonator. In this technique, one of the resonator mirrors is made nonreflective during an interval of pumping of the laser medium. Then the mirror is suddenly made highly reflective. As a result of the change, all the energy stored in the medium during the pumping interval is emitted in a powerful pulse of light, which lasts only for about 10-billionths of a second.

The simplest way of Q-switching a resonator is to rotate one mirror very rapidly. Only during the brief time that it is lined up with the other resonator mirror (at the other end of the resonator) will laser emission occur. Another technique is to place, in the resonator, a dilute solution of a dye that absorbs light at the laser frequency. The absorption of light by the dye initially prevents feedback. This causes an energy buildup in the laser medium. A point will finally be reached when the dye becomes saturated and can hold no more light. When this point is reached, feedback is restored and the laser emits an intense Q-switched pulse of light.

Mode Locking in Amplification Control

During Q-switching with a dye, further intensification of the output beam can be obtained by *mode locking*. In a mode-locked laser, there is a simultaneous oscillation of a number of closely spaced frequencies locked in time in a certain relationship to each other. The spectacular result is an even shorter pulse of only a few trillionths of a second in duration. During this brief instant, however, the laser beam can reach tens of trillions of watts of power. This is more power than is being generated at any given instant from all of the electric power stations in the world.

TYPES OF LASERS

There are a number of lasers available, and each has a different use and purpose.

Insulating Solid Laser

The first prototype laser used ruby as the light-emitting medium. A ruby is a crystalline aluminum oxide containing a small amount of chromium. The chromium atoms are responsible for the red color of the emission. They can be excited by optical pumping with either green or violet light.

The first ruby lasers could be operated only by pulses, (intermittently) because the pumping light had to be extremely intense. If the beam were continuous, the laser could overheat and actually destroy itself. Later designs could be operated at room temperature continuously, rather than by pulses.

The ruby-pulse laser is now used mostly as a high-power pulse laser. Operated in a Q-switch mode and followed by a second ruby laser used as an amplifier to intensify the pulse, a modern ruby laser can deliver pulses of several billion watts of power. These pulses last only several billionths of a second.

Insulating-Crystal Laser

The insulating-crystal laser material typically consists of a hard, transparent crystal doped with a small amount of an element. This is often a rare-earth element that has energy levels suitable for laser emission. A laser built around such a crystal emits in the visible light or the near-infrared region. It often needs to be cooled far below room temperature to operate. Such lasers require intense optical pumping, and they are generally operated only by pulses to avoid overheating. Also, the output frequency of most insulating-crystal lasers can be tuned over a range of a small fraction of a percent by such influences as temperature or magnetic field.

Neodymium is a rare-earth element that is often used as a dopant in various crystals because it has energy levels particularly advantageous for laser action with only relatively modest optical pumping. The best crystalline host for neodymium has been yttrium aluminum garnet (YAG). The characteristic infrared emission from neodymium in this host at room temperature has been obtained on a continuous basis by pumping with a 100 W incandescent lamp, an array of LEDs, or even direct sunlight. Pumping by sunlight could someday provide an ideal possibility for adapting lasers for use on spacecraft.

Glass doped with neodymium atoms is a useful laser material that radiates in the infrared region. Its properties are very similar to those of insulating-crystal lasers. The neodymium-doped glass removes the difficult task of growing large crystals of high optical quality. The glass laser can operate continuously at room temperature, and it has reached efficiencies of about 3% in pulsed operation. Neodymium doped glass lasers have produced the highest output power of any laser: up to tens of trillions of watts. This enormous power was obtained in Q-switched, mode-locked operations in pulses that lasted only a few trillionths of a second.

Semiconductor Lasers

Crystal lasers can also be made from semiconductors such as gallium arsenide (GaAs) and lead telluride. Because these materials can carry an electric current, electronic pumping of semiconductor lasers is possible. A PN junction like those used in transistors is formed in the semiconductor crystal (Fig. 7-3). The junction is put in the forward bias with the positive voltage on the P side and the negative voltage on the N side. Electrons flow through the conduction band into the junction from the N side. Holes flow through the valence band into the junction from the P side. The conduction band is the upper energy

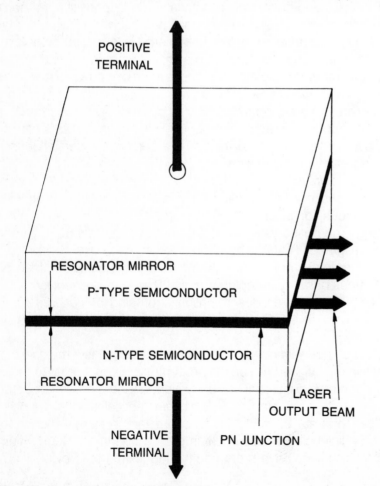

Fig. 7-3. A semiconductor junction laser.

level for the laser, and the valence band is the lower energy level. Thus, an inverted population is established between the upper and lower energy levels. Then laser action occurs. Because the electron and hole flow are referred to as *injection,* these lasers are often referred to as injection lasers.

Injection Lasers

Injection lasers are efficient light sources. They are generally no larger than 1 millimeter (0.04 inch) in any dimension. For the most efficient operation, they must be cooled far below room temperature. For instance, 50% efficiency with a continuous output of several watts has been obtained by cooling a GaAs laser to −253°C (−423°F).

Most injection lasers can operate at very low temperature. However, an important exception is the GaAs laser. This laser can emit infrared light continuously at room temperature with an output of 0.02 W and an efficiency of 7%. It can be made in the form of a heterostructure. In the *heterostructure,* a very narrow PN junction layer of GaAs is sandwiched between layers of a different semiconductor: aluminum gallium arsenide. The properties of this semiconductor confine the electrons and holes to the very narrow junction layer, leading to an inverted population at a lower input current. The laser light is confined to this layer, which makes the resonator very efficient.

Injection lasers can be fabricated to radiate at any wavelength between 0.64 and 32 micrometers by alloying different pairs of semiconductors, such as gallium arsenide and aluminum arsenide. Also, the radiating wavelength of an individual injection laser can be tuned by temperature, pressure, and magnetic fields.

Until recently, the injection laser diode was an exotic and expensive component. Now, it is reasonably inexpensive (some surplus laser diodes can be bought for under ten dollars) and is as readily available as any laser component.

Because an injection laser diode is relatively easy to work with and does not require any special mechanical skills, the projects in this book will use this type of laser diode.

Gas Lasers

The first gas laser used a mixture of helium and neon that was pumped by an electrical discharge. Flat mirrors inside a long glass tube containing the gaseous mixture formed the resonator. In later designs, spherical mirrors placed outside the gas tube were found to make a more convenient arrangement (Fig. 7-4).

The helium atoms are excited by collisions with electrons in the electrical discharge. These collisions create an inverted population

80

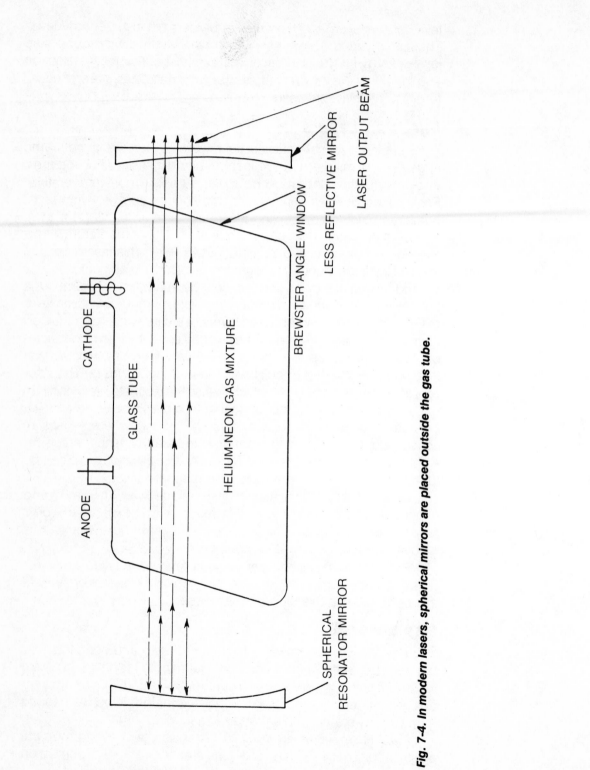

Fig. 7-4. In modern lasers, spherical mirrors are placed outside the gas tube.

in the neon atoms through resonant energy transfer. This energy transfer results in laser emission. The sequence of collisions is a very effective pumping mechanism; however, it cannot always be used because it depends on a coincidence of particular energy levels. Helium and neon atoms provide such levels, but the system might not work for other types of gases.

The original helium-neon laser emitted a beam with a wavelength of 1.15 micrometers, which places it in the near-infrared part of the spectrum. With minor alterations, this laser might also emit radiation in the red part of the spectrum, or at any of several infrared wavelengths.

Another gas laser that also relies on atom-atom collisions uses a mixture of nitrogen, carbon dioxide (CO_2), and helium. In this mixture, the carbon dioxide actually provides the laser emission, so the unit is known as a CO_2 laser.

The CO_2 laser can generate up to 16 kW of power continuously at an infrared wavelength of 10.6 micrometers. If the emission is in the form of very short pulses, this laser can generate several billion watts.

Inverted populations can also be produced by electron-atom or electron-molecule collisions that occur in electrical discharges in certain gases, such as neon, argon, and carbon monoxide (CO). This technique can also be used with vapors, such as mercury vapor or water vapor. Hundreds of gas laser designs use this type of collision pumping.

In another type of collision pumping, a high-energy electron beam is used to irradiate a gas. This method has been used to obtain laser emissions of ultraviolet waves from hydrogen, xenon, and krypton.

In still other gas lasers, a dissociation of molecules is obtained in the pumping process either by electron-molecule collisions or by the absorption of light energy. Oxygen, chlorine, and methyl iodine are some of the gases used in these lasers.

Ionized atoms of a gas or vapor can also be used to produce a laser emission. A gas in this ionized state is called a *plasma*. Argon, cadmium, and neon have all been used in this manner.

Gas Dynamic Laser

Sometimes large engines force gases through the resonator at a supersonic velocity. The supersonic-velocity gases remove the excess heat generated in the electrical discharge used for the excitation. Therefore, they permit a higher-power, continuous output. A laser that uses gases at supersonic velocities is called a *gas dynamic laser*.

The gas-dynamic laser can also be used to produce gas-expansion pumping. If a uniformly hot mixture of nitrogen, CO_2, and water vapor is expanded at supersonic speed, the nitrogen pumps the upper laser energy level in CO_2 by collisions. This results in an inverted population and subsequent laser action. Continuous power outputs of up to 60 kW have been achieved in this manner.

Chemical Lasers

In a chemical laser, a chemical reaction creates and pumps gas. The chemical pumping occurs through the release of energy in an exothermic chemical reaction. An example of this process is the reaction of hydrogen and fluorine. The reaction produces hydrogen-fluoride molecules with an inverted population, which leads to laser action.

Usually, chemical lasers use the gas-dynamic configuration so that the de-excited chemical reaction products and excess heat are quickly removed from the resonator.

Liquid Lasers

The most important class of liquid lasers uses a dilute solution of an organic dye in an organic solvent. Scores of dyes have proven useful. One of the most prominent is Rhodamine 6G. Laser emissions throughout the region from near-ultraviolet (0.32 micrometer wavelength) to near-infrared (1.2 micrometer wavelength) have been obtained by using various dyes. All are optically pumped, often with another laser. Several of these laser types can be run continuously, and others can be mode-locked to produce pulses that have a duration of only a few trillionths of a second.

The most remarkable and useful property of a dye laser is the large and continuous tunability of its laser wavelength. In a single solution of slightly acidic 4-methylum belliferone, the laser emission has been tuned from a wavelength of 0.391 micrometer in the near-ultraviolet region to 0.567 micrometer in the yellow-green region.

As you can see, there is a broad range of materials and substances available that have lasing or masing capabilities. The techniques at your disposal affect such laser and maser emissions. If you have an LED wrist watch, you probably have not been aware that you were carrying a near-laser device. When you see a common neon display sign in a store window, you are witnessing the preliminary laser functions. These examples are just two of the many everyday items that make up the laser-parts warehouse of ideas.

Chapter 8

Phasors

IN A PREVIOUS CHAPTER, YOU LEARNED THAT PHASORS CAN BE USED TO arrive at resultant values of electromagnetic waves. Bear in mind that to use phasors for that purpose, all the waves in question must be at exactly the same frequency. For example, the resultant wave for *all* waves of red light (with a wavelength of 6328 nanometers) can be resolved, but all other wavelengths must be resolved separately.

Of specific interest is the effective power, or equivalent balance, of many different electromagnetic waves occurring simultaneously. These are the many stimulated photon emissions within the laser or maser.

PARALLELOGRAM

Waves, even those of the same frequency and wavelength, quite often appear out of phase with each other. In other words, there can be some occurring 10° apart in time, some 60° apart in time, and others occurring at another phase difference. Remember that all the waves evaluated are to be of the same frequency. The only differences are the phases and the amplitudes. Actually, the resultant effects of all the total characteristics of the waves will be collected. For the waves in Fig. 8-1, the arrows or measured lines (f-1 and f-2 in Fig. 8-2) are

84

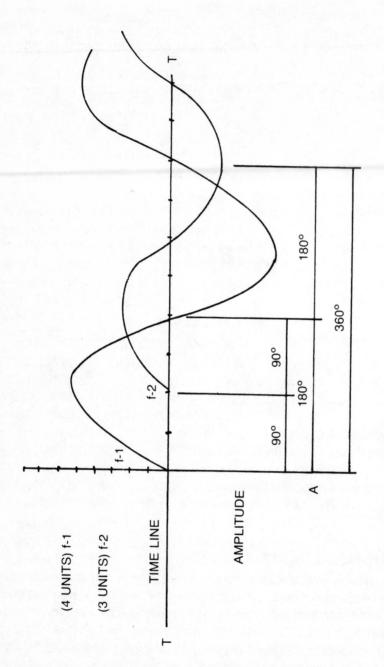

Fig. 8-1. Ordinates on the time line of a wave graph.

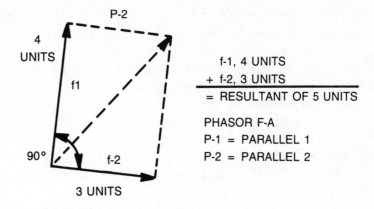

f-1, 4 UNITS
+ f-2, 3 UNITS
= RESULTANT OF 5 UNITS

PHASOR F-A
P-1 = PARALLEL 1
P-2 = PARALLEL 2

Fig. 8-2. Resultant of a phasor parallelogram.

represented as adjacent sides of a parallelogram. For every set of component waves, the arrows are drawn with exactly proportional lengths. Thus, for two components with lengths of 3 units and 4 units, the parallelogram would have sides measuring 3 inches and 4 inches. The corner angle would be determined by the phase difference between the two waves. A diagonal measured across the interior of the parallelogram is exactly 5 inches long. This is not an algebraic addition; it is a type of addition that considers the effect of phase difference.

In constructing a phasor, first select a starting point, usually the first component off of the time line. Then construct a parallelogram for this component, using it and the component adjacent to it. (Place the second component in a position that is counterclockwise from the first component.) The length of each line is proportional to the amplitude of the represented component. If there are just two components, the resultant, or phasor, measured diagonally across the interior of the parallelogram directly represents the effective combination of the two components.

In many situations, there will be more than two components to consider. In such a case, first find the phasor for the first two components as described in the preceding paragraphs. Then use that resultant phasor and a new line representing the next component wave (continuing to move counterclockwise) as the sides of the next parallelogram. Repeat the process until all components have been included. In place of all of the individual components, you now have a single phasor equivalent (f-C in Fig. 8-3). This represents the effect of the total of all the individual components if they were to have a chance to act independently.

Try a specific example. As shown in Fig. 8-1, construct the ordinates. The horizontal line represents time. The time line is marked off in degrees—360 units equalling one complete cycle.

Using this time line as the reference, determine the relationship between adjacent waves. Figure 8-1 shows that wave f-1 *leads* wave f-2 by 90° because wave f-1 reaches maximum amplitude 90° closer to zero on the time line. Conversely, wave f-2 *lags* wave f-1 by 90° because wave f-2 reaches maximum amplitude 90° farther from zero on the time line. The only difference in the two ways of stating the relationship is that a different wave is selected as the reference.

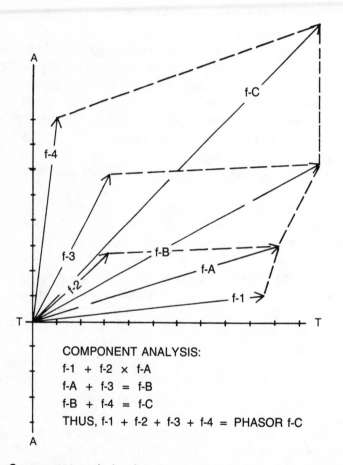

COMPONENT ANALYSIS:
f-1 + f-2 × f-A
f-A + f-3 = f-B
f-B + f-4 = f-C
THUS, f-1 + f-2 + f-3 + f-4 = PHASOR f-C

Fig. 8-3. Component analysis of a phasor diagram. The vertical sum of all waves, or values, equals the equivalent of one wave having the effective value of f-C.

COHERENCY

The coherency of laser light, meaning the in-phase characteristic of laser light, permits us to apply such a phasor analysis to determine the effectiveness or resultant power of such an electromagnetic wave. Ordinary incoherent light, by virtue of its multiplicity of wavelengths would make this application impossible. The main body of laser emission is coherent and of the same frequency, differing only slightly in its exact phase. It keeps pretty near to the absolute 0° phase as the inherent method of generation. Stimulated photon emission assures all photon emissions to occur at the same frequency.

Figure 8-3 illustrates the vectorial system in which a polyphase analysis is made of four separate values. The diagonal (resultant) of the first completed diagram serves as one side of the next parallelogram to be constructed. This progression is duplicated for each and every component to be vectorially collected.

Chapter 9

Introduction
to the Projects

SO FAR IN THIS BOOK, YOU HAVE STUDIED LASERS ON A THEORETICAL LEV-el. Next, you can get some hands-on experience. The remaining chapters are devoted to practical projects for you to build and experiment with.

Some of the projects involve simulated rather than true lasers, because simulated lasers are less expensive, easier to work with, and safer than the real thing. The simulated laser project of Chapter 10 is strongly recommended as a first project for anyone who does not have practical experience with lasers.

When actual lasers are used in the projects, they will be of the injection-diode type. This is for economy and ease of construction. Also, laser diodes seem to be somewhat more readily available than other laser devices. Suitable injection diodes are featured in many surplus and retail catalogs. Shop around for the best deal.

The most commonly available injection diodes emit infrared light. Because infrared light is not visible, you will need to build the detector described in Chapter 11 to detect laser output.

PROJECT REQUIREMENTS

Esoteric or hard-to-find parts are avoided in the projects. However, specialized items are required in some cases, and wherever possible, alternatives are suggested.

In most cases, you will need one or more lenses for the projects, so you should get a catalog from a dealer of optical devices. The lenses used in these projects will be fairly simple types, and you should have no difficulty finding a suitable lens.

If you regularly build electronic projects, you should have most, if not all, of the tools you will need to construct the projects in this book. At a minimum, have the following tools available:

- soldering iron
- pliers
- diagonal cutters
- artist's or craft knife
- voltmeter

Other useful items include an oscilloscope and one or more small vises or clamps.

To avoid potential difficulties, step-by-step instructions are given for each of the projects. If you work slowly and carefully, and follow all instructions, you should have no problem constructing any of the projects in this book.

SAFETY PRECAUTIONS

There is always some potential for danger whenever a laser is used. The concentrated, high-energy output of a laser is potentially very destructive. The projects presented here are for low-power lasers only, but there is still some risk, and precautionary measures must be observed. Of course, the laser simulator is not a true laser, and it should be as safe as any electronic circuit.

When working with a true laser, take extra caution if children or pets might be around. Always stop and think about whether anyone else might come into the range of the laser beam. Take precautions to be sure that no one wanders into the area while you are operating the laser. It is better to be overcautious than to have a tragic and unnecessary accident. Do not get careless just because you are wearing safety goggles.

Observe the following safety precautions when working with, experimenting with, or operating any laser or other radiation-emitting device. These precautions are recommended by The Bureau of Radiological Health, Division of Compliance, Rockville, Maryland 20852.

• Never permit the eye to be in the direct path of the laser beam, no matter how low the power. Never operate a laser in any manner that would permit others to unknowingly look directly into the beam. This is especially important when the beam is invisible (such as with an infrared laser). A laser beam, whether visible, infrared, or ultraviolet can cause serious burning of the retina of the eye. It bears repeating: never look directly into any laser beam, even briefly! Permanent damage can be done very quickly.

• To be safe, always avoid looking into the expected laser beam path when there is any possibility that the laser might be fired, even accidentally.

• It is a good idea to wear protective goggles. Make sure the goggles are adequate for the expected radiation outputs. Goggles may not be necessary for work on low-power lasers like the one in this book, but they are still a good idea, just to be safe.

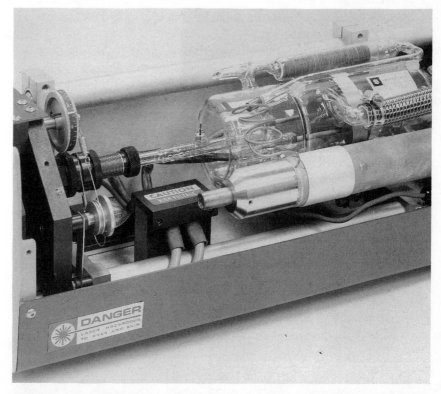

Fig. 9-1. Post safety signs not only around the area of experimentation but also on the laser itself.
Photograph courtesy of Spectra-Physics.

- Even with goggles, never permit the eye to intercept the direct path of the laser beam, the beam landing, alignment, or projection.
- Keep all specular reflectors and mirrors out of the path of the laser beam (except, of course, when such reflectors are a controlled part of a project).
- Avoid work environments with extremely low ambient illumination.
- Keep all laser apparatus completely enclosed wherever possible. If the lasers are operated without covering, then additional radiation may be reflected.
- Terminate the laser beam at the target by providing a backstop of carbon blocks.
- Never permit a laser to be left unattended while it is operating.
- Always make a neon indicating lamp an integral part of every laser operating circuit. This applies to all circuits given in this book. This precaution is needed because infrared and ultraviolet emissions are invisible to you, and could exist without your knowledge.
- Post signs to identify the operating area as an area of radiation or a laser test area. A typical sign is shown in Fig. 9-1. Take care that persons wearing or requiring heart pacemakers are not exposed to such radiations.

These precautions are not intended to frighten you. Low-power laser projects like the ones in the following chapters are not particularly dangerous if you are careful.

Chapter 10

Simulated Laser Project

THIS FIRST PROJECT IS NOT A TRUE LASER. IT MERELY GIVES A CRUDE SIM-ulation of laser action. You might be tempted to skip this project and get on to the real thing, but you are very strongly urged to build this project first. Even if you have considerable experience with electronics, you will find that lasers are very specialized and complex devices. Building a simulator is a good way for you to familiarize yourself with the basics before you launch into a complex and potentially dangerous project like a working laser. The simulator will also come in handy as a testing and calibration device for the other projects in this book. The simulator emits a beam of visible light. The semiconductor laser (presented in Chapter 12) emits only invisible light, which can obviously be tricky to work with because you cannot see it.

A true laser can be quite dangerous if used improperly or if it is carelessly constructed. This simulator project is as safe as an ordinary flashlight. In addition to being an educational demonstration device and convenient testing device, this project can also be a safe toy for children. It can also be used as the centerpiece of a fine science-fair project.

Lasers operate in one of two modes: continuous or pulsed. A continuous laser produces a steady, unvarying beam. A pulsed laser emits bursts of light; the beam is turned on and off at short intervals. The simulated laser project is equipped with a switch that allows you to select either of these two modes of operation.

This project uses an ordinary LED as the light source, in place of a true laser diode. Actually, LEDs and laser diodes are quite similar in their internal construction. In a sense, a laser diode is a specialized LED.

Two tricks are used to make a simple LED simulate a laser. First, the LED must be overdriven; that is, more than the normal amount of current is fed through the LED so it glows more brightly. This may shorten the life of the LED somewhat, but there should not be any noticeable problem. Use the largest and brightest LED you can find.

The second trick is to use a lens to focus the omnidirectional light from the LED into a narrow beam. This beam will not be coherent. The beam will spread out far more than the beam from a true laser, but it is a reasonable simulation.

The schematic diagram for the laser simulator project is shown in Fig. 10-1, and the parts list is in Table 10-1. As you can see, there is nothing very complex about this project.

Use four AA cell batteries for a power supply voltage of 6V (nominal). Switch S1 is a SPST (single-pole, single-throw) power switch to turn the circuit on and off. Switch S2 is a SPDT (single-pole, double-throw) switch used to select the operating mode. This switch determines how power will be fed to the LED.

For continuous operation, power is fed directly to the LED from the batteries through resistor R3. Refer to Fig. 10-2 for a simplified equivalent circuit of the continuous mode. This is a simple circuit to light an LED. Resistor R3 is a current-dropping resistor. The only thing special in this circuit is the value of the current-dropping resistor.

A current-dropping resistor must always be used with an LED. Without this resistor, the LED would draw excessive current and burn itself out. Normally, this resistor would have a value between 200 and 1,000 Ω. In this application, use a much smaller resistance (68 Ω). The smaller this resistance is, the brighter the LED will glow. Recall that the LED is being overdriven in this project. You should not have a problem with LEDs burning out, but if you do, increase the resistance of R3 slightly. You might also get a little more light output by reducing the resistance, but be careful because you might damage the LED if you use a resistor that is too small.

In the pulsed mode, the LED is driven from the output of a simple astable multivibrator circuit built around a 555 timer. When the multivibrator output is low, the LED will be off. When the multivibrator output goes high, the LED glows. The multivibrator signal is fed to

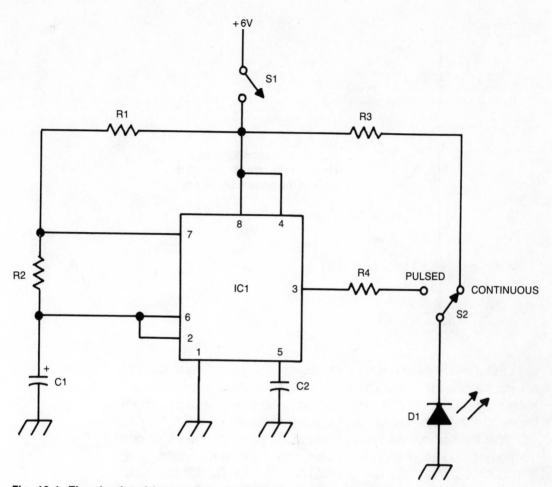

Fig. 10-1. The simulated laser safely simulates the action of a laser.

Table 10-1. Parts List for the Simulated Laser of Fig. 10-1.

COMPONENT NUMBER	COMPONENT DESCRIPTION
IC1	555 timer
D1	LED
R1	10 kΩ resistor
R2	100 kΩ resistor
R3, R4	68 Ω resistor
C1	10 μF, 15V electrolytic capacitor
S1	SPST switch
S2	SPDT switch

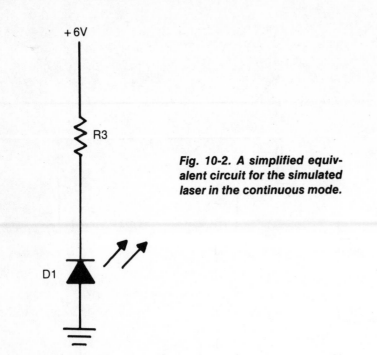

Fig. 10-2. A simplified equivalent circuit for the simulated
laser in the continuous mode.

the LED through current-dropping resistor R4. In the parts list, this resistor has the same value as R3, but you might want to use different resistance values for the continuous and pulsed modes. Slightly more brightness might be useful for the pulsed mode.

With the component values listed in the parts list, the pulse rate will be about ¾ second on and ¾ second off, for a complete cycle of about 1½ seconds. Experiment with other values. Remember, however, that if the pulse rate is made too high, your eye will not be able to distinguish between the individual pulses. The simulated laser will appear to be operating continuously, although perhaps at slightly lower brightness.

It is always a good idea to breadboard a project before permanently constructing it. Breadboarding will allow you to test and modify the circuit (if you choose), and it will give you a better feel for what is happening.

If you breadboard the circuit, you probably will not be too impressed. As described so far, the circuit does not resemble a laser very closely. You just have an LED that can be made to either glow continuously or blink on and off at a slightly higher than normal intensity.

The first trick required for this project is overdriving the LED; this is accomplished through selection of resistance values as already de-scribed. The second trick you must use is to focus the light emitted by the LED. You can focus LED output by placing the LED at one end of a telescopelike device with a lens at the other end.

You will need two pieces of tubing, either plastic or metal. (Plastic will be easier to work with.) Each tube should be about 1 foot long. For convenience, refer to them as tube A and tube B. The inside diameter of tube A should be equal to the outer diameter of tube B. In other words, tube B should fit easily, but snugly inside of tube A. You should be able to slide tube B in and out of tube A easily in telescope fashion. The tubes should fit together as shown in Fig. 10-3. Be sure to check and remove all burrs or irregularities on the edges of the tubes. (This is especially important if metal tubing is used.) A small burr on one tube could catch on the other tube and prevent easy movement of the tubes.

You should be able to find suitable tubing at a large hardware or building supply store. Diameters of approximately 1.5 inch will do nicely.

For best results, spray paint the interior of both tubes with flat black paint.

Next you will need a convex lens to fit the end of the larger tube (tube A). The exact requirements of this lens are not too critical. You can find many suitable lenses at an optical supplier. You might be able to get by with the lens from a low-power, toy magnifying glass.

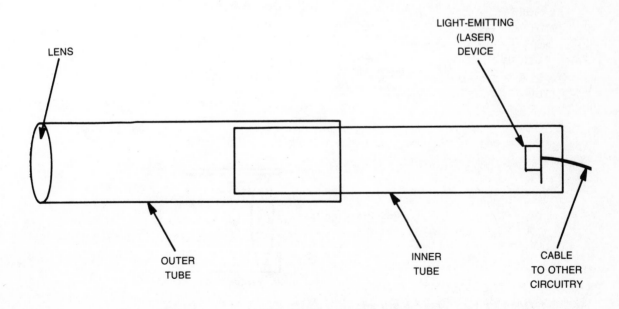

Fig. 10-3. A pair of telescoping tubes are used to focus the beam.

Experiment with several different lenses until you find one that works well. Almost any convex lens will work. The characteristics of the lens used will determine the focal point of the device.

Probably the hardest part of this project is mounting the lens. There are many possible methods. The best one will depend on your personal preference and the materials you have available. Use your ingenuity.

One method of mounting is to use two plastic end caps as shown in Fig. 10-4. One of the caps should fit snugly inside the larger tube. The other should fit over the outside of the top. Using an artist's or craft knife (or similar tool), cut an opening in the end of each of the

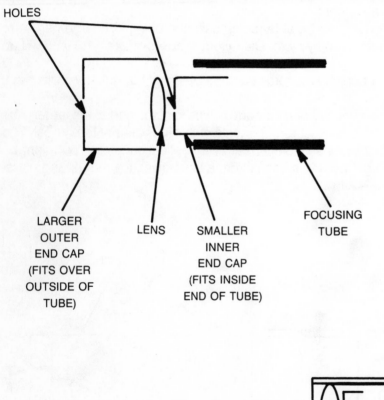

HOLES

LARGER
OUTER
END CAP
(FITS OVER
OUTSIDE OF
TUBE)

LENS

SMALLER
INNER
END CAP
(FITS INSIDE
END OF TUBE)

FOCUSING
TUBE

EXPLODED VIEW

ASSEMBLED

Fig. 10-4. One method of mounting the lens.

caps. Fit the smaller cap into tube A, position the lens over the end of the tube, and place the second (slightly larger) end cap over it.

If you use glue in the construction of this project, be very, very careful to avoid getting any on the lens. Even if the glue dries clear, a small smear can significantly distort the optical characteristics of the lens.

Build the laser simulator project in two parts, with a length of two-conductor cable connecting them. The finished project is shown in Fig. 10-5.

Most of the circuit is constructed on a small piece of perforated circuit board, as shown in Fig. 10-6. You could make up a customized printed circuit board if you prefer. In either case, all of the components shown in the schematic diagram (Fig. 10-1) are mounted on this board, except for the two switches and the LED. The board is housed in a small minibox, as shown in Fig. 10-5. The two switches are mounted on this box. The batteries and holder are housed in this box. In place of the LED, one end of the connecting cable (two-conductor) is wired into the circuit. The other end of this cable is fed out through a small hole in the box.

The LED is mounted on a small piece of perforated circuit board cut to fit snugly inside tube B (the smaller tube) at the far end from

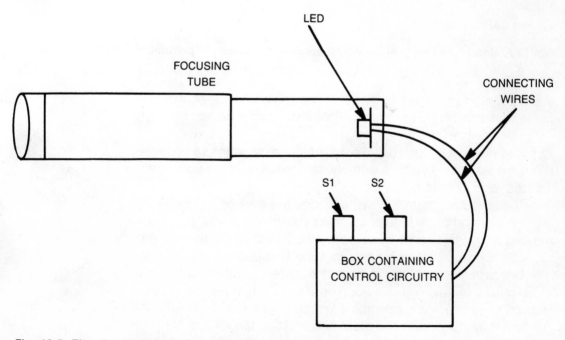

Fig. 10-5. The simulated laser is built in two parts.

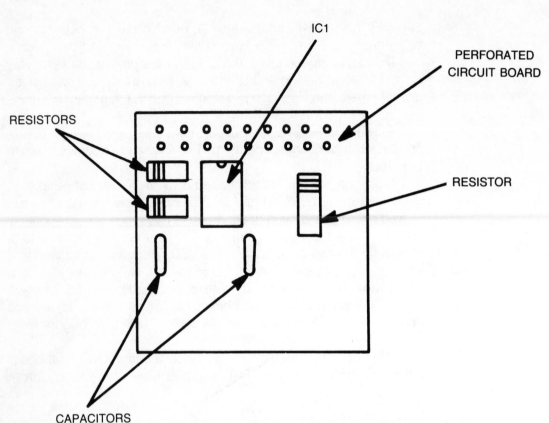

Fig. 10-6. Most of the simulator circuitry is constructed on a small perforated circuit board.

the lens. This is illustrated in Fig. 10-7. The free end of the connecting cable is soldered to the LED leads. (Be sure to use the correct polarity!)

Some electrician's tape, or glue might be needed to hold the LED board in place. It is best to avoid a permanent attachment because the same telescoping tube assembly can be used for the solid-state, infrared laser project of Chapter 12.

The project is complete. It will work best in a dark or dimly lit room. Aim the lens end of the tube at whatever you want to fire at, and focus the beam to a small spot by adjusting the length of the telescoping tube. Focus is controlled by how far tube B is pushed into tube A.

Because this is not a true laser, it is perfectly safe. You can aim the simulator at anything you like. There is no risk to people or pets. You can even look directly into the beam, if you like. The beam produced by this circuit is no more powerful (or dangerous) than the beam from an ordinary flashlight. (Cats love to chase the beam spot around as you move it across the floor.)

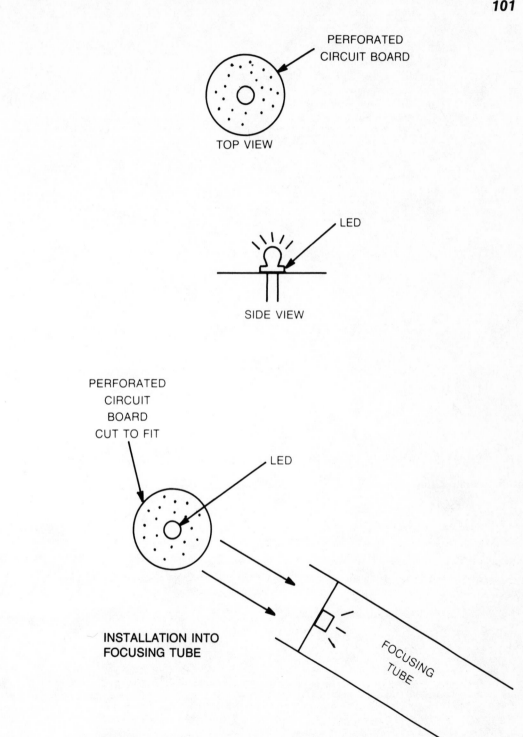

PERFORATED
CIRCUIT BOARD

TOP VIEW

LED

SIDE VIEW

PERFORATED
CIRCUIT
BOARD
CUT TO FIT

LED

INSTALLATION INTO
FOCUSING TUBE

FOCUSING
TUBE

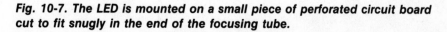

Fig. 10-7. The LED is mounted on a small piece of perforated circuit board cut to fit snugly in the end of the focusing tube.

Chapter 11

Infrared Detector Project

IN CHAPTER 12, YOU WILL BUILD AN ACTUAL LASER. THE BEAM FROM THIS laser is in the infrared region, so the light is invisible. Because you cannot see the beam, you need some other way to determine its presence. This detector gives you the means to sense the invisible beam.

The detector gives a two-fold indication when it detects any infrared light: an LED lights up, and a tone sounds. You can easily eliminate either the tone or the LED, if you choose. The schematic diagram for the infrared detector project appears in Fig. 11-1, and the parts list is in Table 11-1. Refer to the block diagram in Fig. 11-2. The circuit is very simple.

The detector is a special phototransistor designed to be sensitive to light in the infrared region. In some cases, performance can be improved by placing a filter over the sensor. A sheet of deep red plastic will do.

As the amount of infrared light reaching the sensor varies, the effective resistance of the phototransistor Q1 changes. The phototransistor and resistor R1 act as a simple voltage divider. The actual voltage at the junction of R1 and the collector of Q1 will depend on the resistance between the collector and emitter of Q1. This resis-

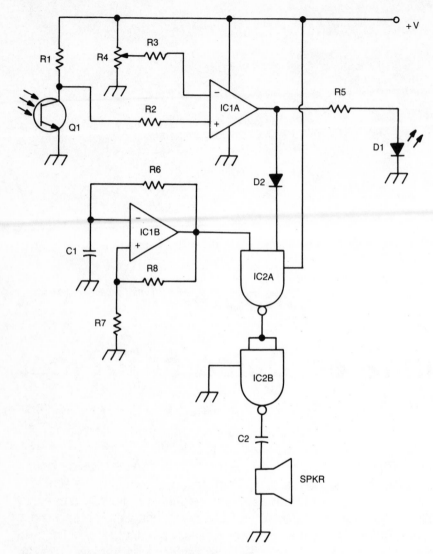

Fig. 11-1. This circuit detects infrared light.

tance is controlled by the amount of infrared light striking the sensor (the base signal).

The R1 / Q1 voltage is fed to the input of a voltage comparator (IC1A and its associated components). A 324 quad operational amplifier is shown in the parts list, even though only two sections are required. A pair of 741s, or a 747 dual operational amplifier chip could be used instead, but those devices require a dual polarity power supply. With the 324, you can use a single-ended supply, so the total circuit cost is lower. For this application, there is no need for a high-grade operational amplifier.

Table 11-1. Parts List for the Infrared Detector of Fig. 11-1.

COMPONENT NUMBER	COMPONENT DESCRIPTION
IC1	324 quad operational amplifier (see text)
IC2	CD4011 quad NAND gate
Q1	infrared sensitive phototransistor
D1	LED
D2	1N4001 diode
R1	100 Ω resistor
R2, R3	1 kΩ resistor
R4	10 kΩ trimmer potentiometer
R5	330 Ω resistor
R6	47 kΩ resistor
R7	10 kΩ resistor
R8	100 kΩ resistor
C1, C2	0.0 μF capacitor
SPKR	small speaker

Be only concerned with whether or not the R1 / Q1 voltage is larger than the reference voltage, which is set by potentiometer R4. A potentiometer is used to permit adjustment of circuit sensitivity.

When a sufficient infrared signal is detected, the output of the comparator goes high (to a value near the supply voltage). The comparator output is split into two paths. In one path, it is fed through current-limiting resistor R5 to an LED. When an infrared signal is detected, the LED will light up, providing a visual indication. When no infrared signal is received by Q1 (or the infrared level is too low to create enough voltage to trigger the comparator), the LED will be dark.

The other path for comparator output signal is through a simple gating network (IC2A and IC2B). These two NAND gates are combined to create an AND gate. (IC2B is wired as a NOT gate, or inverter.) A single AND gate could be used, but NAND gates tend to be more readily available and less expensive. Because there are four gates on a single integrated circuit, this two-step method does not add to circuit bulk or cost.

The other input to the gate is a square-wave signal produced by IC1B and its associated components. Once again, an inexpensive operational amplifier is fine. This is another section of the 324 quad

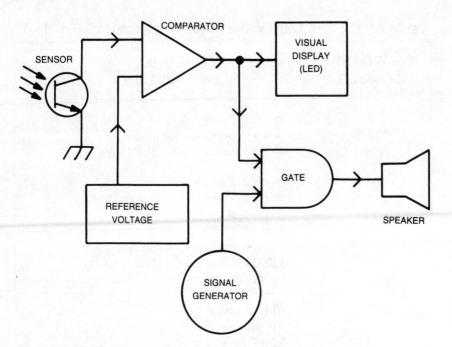

Fig. 11-2. Block diagram of the infrared detector project.

operational amplifier. The remaining two sections are not used in this project.

Amplifier IC1B is connected as a simple square-wave generator. With the component values listed in the parts list, the frequency will be approximately 1,000 Hz. This provides an audible test tone.

To understand how the audible signal is controlled, you need to know how an AND gate works. There are two inputs; each may be either low or high. There are no other possibilities. The single output also can be either low or high, depending on the input combination. At any given instant, there are only four possible input combinations. For convenience, call the tone signal A and the comparator signal B. The four possible combinations and the resulting outputs are as follows:

A = low	B = low	Output = low
A = low	B = high	Output = low
A = high	B = low	Output = low
A = high	B = high	Output = high

The output of the gate is high if, and only if, both input A and input B are high. If either, or both, of the inputs is low, then the output is low.

The tone signal (A) continuously switches back and forth between the low and high states, as long as power is applied to the circuit. The comparator signal (B) goes high when an infrared light source is detected. If there is no infrared light source, then input B is low.

When input B is low, the gate output will be low, regardless of the state of input A. As long as no infrared light is detected, no tone signal can get through the gate to the speaker; the speaker remains silent. When an infrared signal is detected, the comparator signal (B) goes high, and gate output is controlled by the tone signal (A). The gate output switches back and forth between states in step with the tone signal. This fluctuating signal is fed to the speaker, and a tone indicates the presence of infrared light.

Depending on your personal preference, you might want to add a switch to cut off the tone generator portion, or you might even consider eliminating this portion of the circuit completely.

Nothing in this circuit is critical. You should not have problems even if you decide to substitute other component values.

When you make a case for this project, cut a hole in the case as the infrared beam target (see Fig. 11-3). Carefully position the phototransistor directly behind this hole. If the sensor is not positioned properly, the infrared light will not be able to reach it. There is a simple way to check for correct positioning. Do you have a clear, direct view

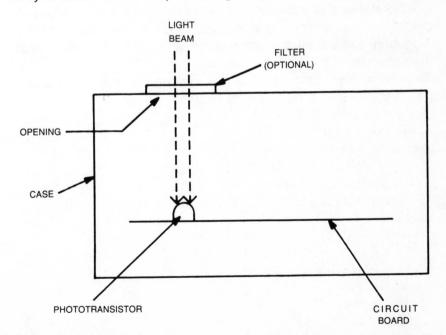

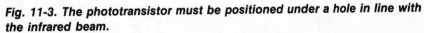

Fig. 11-3. The phototransistor must be positioned under a hole in line with the infrared beam.

of the sensor surface through the hole? If you cannot see it, the infrared light beam cannot reach it.

Cover the opening with clear plastic or a filter to protect the sensor and the circuitry. Use a red filter or a diffusing filter. A diffusing filter can be cut out of a white, translucent, plastic milk carton. If a diffusing filter is used, aiming the infrared beam at the sensor will be slightly less critical. You can use a larger opening with a diffusing filter.

Testing the infrared detector is not difficult. If you are reading a book like this, it is a safe assumption that you like electronic gadgets. You probably have at least one wireless remote control. It does not matter what it is for: a video cassette recorder, a television, or a compact disc player. If you do not have a wireless remote control, borrow one from a friend.

Most modern wireless remote controls use infrared pulses. When a control button on the remote controller is depressed, an infrared signal is emitted. To check your infrared detector, just hold the remote control near the sensor opening, aim it carefully, and press a button for any function. The detector should indicate that it has received a signal by lighting the LED and sounding a tone. For many remote controls, the infrared beam will be emitted as a very brief pulse, so pay attention to the LED as you push the buttons. (The audible tone of the speaker will be hard to miss.)

Carefully adjust potentiometer R4 for the desired sensitivity. The detector should indicate the received beam reliably without false triggering. Be aware that in some cases, ambient lighting may affect the sensitivity of this circuit.

This is a very simple project, but it will be useful in working with the laser projects in the following chapters.

Chapter 12

Continuous-Wave, Solid-State Laser

FINALLY, YOU CAN WORK WITH AN ACTUAL LASER. BEFORE GETTING START-
ed on this project, you are very strongly advised to reread the safety
precautions in Chapter 9. An additional review of safety precautions
is given in the following paragraphs.

Never look directly into a laser beam. Do not focus the beam on
an open mirror. Never fire the laser at anyone's face or anywhere
where someone might accidentally look into the beam.

Because the laser device described here produces an invisible
infrared beam, it is especially important for the operator to be extra
careful. The beam is invisible, and someone could look directly into
it, without even realizing it until it was too late.

Even though this is a low-power laser project, the beam is intensive
enough to cause permanent eye damage if the laser is used carelessly.

Use particular care when experimenting with lasers if there are
any children or pets around. Unless you are very closely controlling
a demonstration with all required safety precautions, lock children and
pets out of the work area whenever you activate the laser. If anyone
else (especially a child) might have access to your equipment, disable
the laser whenever it is unattended. You could disconnect the power

WARNING!

This laser device emits invisible infrared radiation.
Potentially hazardous if eyes or skin exposed to laser beam.

USE WITH CARE!

WARNING!

Fig. 12-1. Attach a warning sticker like this to your project.

supply, or use a key switch to turn the equipment off. Keep the key in a secure place.

Make up some warning stickers like the one shown in Fig. 12-1, and attach them to any laser projects you build. Do not omit the warning labels, or you might cause serious injury and be legally responsible for the accident.

If you are careful, you should not have any problems with this or other projects in this book. But if you are careless, the consequences could be very serious, possibly tragic. Please be careful.

THE LASER DIODE

The heart of this laser project is a special semiconductor diode shown in Fig. 12-2. Laser diodes are small and inexpensive when compared to other types of laser devices, such as a helium-neon tube or a ruby rod. For low-power work, a laser diode is generally easier to work with. These components are becoming readily available to you, especially at surplus houses. For these reasons, all the laser projects in this book use laser diodes.

The technical name for the laser diode is the *injection laser*. This device is very closely related to the common LED. An LED is a two-lead, semiconductor device. When the voltage applied across the two leads is below a specific threshold level (determined by the construction of the individual component), nothing happens. When the applied voltage exceeds the threshold level, the semiconductor junction starts to emit light—the LED glows.

An injection laser diode has a second critical threshold point, called J_{TH} or I_{TH} (these two terms are interchangeable). When the applied current is below this threshold, the diode functions exactly like an

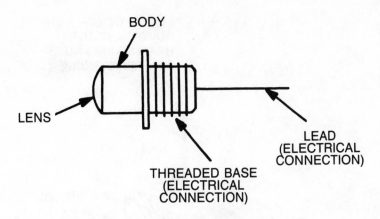

BODY

LENS

LEAD
(ELECTRICAL
CONNECTION)

THREADED BASE
(ELECTRICAL
CONNECTION)

Fig. 12-2. A special diode generates the laser beam.

ordinary LED. That is, the junction glows with a relatively broad spectrum of wavelengths; the light is emitted in a wide pattern of radiation. Under these conditions, the diode does not function as a laser device.

But if the applied current to the injection laser exceeds the threshold level (J_{TH}), the emitted light thins down to a narrow beam. The emitted light is coherent and of a single wavelength. The emitted laser beam escapes from both of the laser end faces (unless one end is coated with a reflective—usually gold—film).

For a clearer idea of just how closely the injection laser is related to the LED, refer to Figs. 12-3 and 12-4. The structure of a typical LED is illustrated in Fig. 12-3, and Fig. 12-4 shows the simplified structure of a diffused junction laser made from a similar semiconductor wafer. The differences in function are illustrated in Fig. 12-5.

CONTACT WIRE

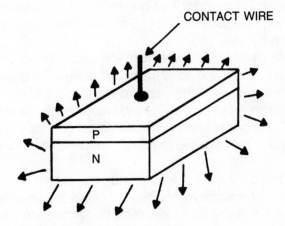

P

N

Fig. 12-3. Physical structure of a typical LED.

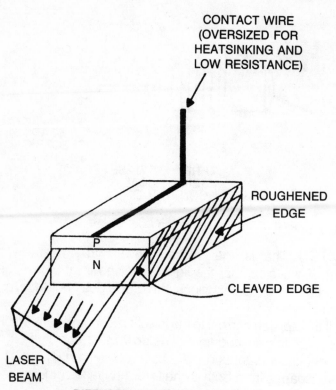

CONTACT WIRE
(OVERSIZED FOR
HEATSINKING AND
LOW RESISTANCE)

ROUGHENED
EDGE

P

N

CLEAVED EDGE

LASER
BEAM

DIFFUSED JUNCTION LASER

Fig. 12-4. An injection laser diode has a physical structure quite similar to that of an LED.

There are a variety of laser diodes that will work for this project. The diode used for the prototype of this project is an LD60. It is a heterojunction GaAs diode. This particular diode was manufactured by Laser Diode Laboratories, Inc. Similar diodes are manufactured by other companies.

The LD60 features relatively high efficiency at relatively low drive currents; these qualities simplify the circuit design. Diodes in the LD60 series (there are several) are capable of up to 25 W of peak output power. The diode is hermetically sealed in an optically centered, coaxial package. The package features an 8-32 screw stud for easy mounting. This stud is also the negative terminal. The positive terminal is a long lead.

The LD60 is fairly easy to use and has few serious restrictions. The reverse voltage across the diode should be kept below 3V. A larger reverse voltage could degrade performance or even cause damage.

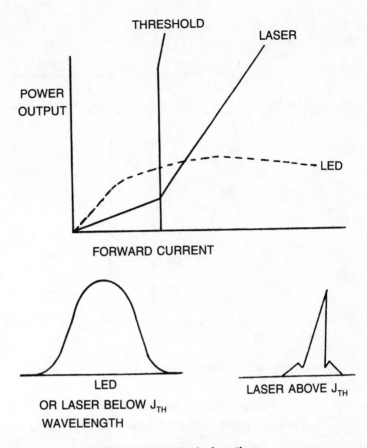

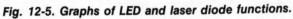

Fig. 12-5. Graphs of LED and laser diode functions.

The biggest problem with any semiconductor is thermal buildup, which can cause the diode junction to self-destruct. The LD60 is rated for operation up to 75° C (167° F). The manufacturer recommends that this unit be operated only by short pulses. It is very easy to overheat and damage the laser diode in a continuous wave circuit like this one. Use a normally open, push-button power switch. Never depress the button for more than a second or so continuously. If power is fed uninterrupted through the laser diode for too long, the diode will almost certainly overheat and be destroyed.

THE CIRCUIT

The schematic diagram for this project is shown in Fig. 12-6, and the parts list is in Table 12-1. This is an experimental circuit; you might have to make minor modifications to allow for differences in component tolerances.

114

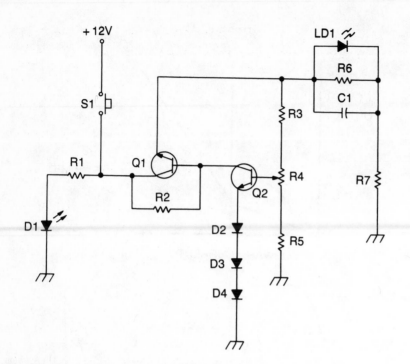

Fig. 12-6. Continuous-wave, infrared laser.

Table 12-1. Parts List for the
Continuous-Wave, Infrared Laser of Fig. 12-6.

COMPONENT NUMBER	COMPONENT DESCRIPTION
Q1	npn power transistor (Radio Shack RS2041, or similar)
Q2	npn power transistor (Radio Shack RS2020, or similar)
D1	LED
D2, D3, D4	1N4003 diode
LD1	Laser diode
C1	120 pF capacitor
R1	330 Ω resistor
R2	100 Ω resistor
R3, R5	1.5 kΩ resistor
R4	5 kΩ trimmer potentiometer
R6	1 kΩ resistor
R7	27 Ω resistor

In the early stages of building and experimenting with this circuit, replace the laser diode with a 1N4005 diode. This diode will safely imitate the electrical characteristics of the laser diode. Make sure that everything in the circuit is working properly before installing the laser diode. There are two reasons for this. The first reason is that the laser diode is the most expensive and delicate component in the circuit. There is no sense in subjecting it to unnecessary risk. The second reason is that using a 1N4005 diode in place of the laser diode will protect you. You will not need to worry about shielding the laser beam. You can safely work on the circuit in the open.

This circuit is a high-gain current amplifier. Considerable current is drawn from the power supply. If batteries are used, they must be able to supply a fairly high current. Rechargeable ni-cad (nickel cadmium) cells or an ac power supply is your best choice.

Because the emitted laser beam is invisible, an indicator device is essential so that you will know when the circuit is in operation. This is the function of diode D1. This is a common LED, positioned where it is conveniently visible to you. Resistors R1 and R2 make up a voltage divider to drop the 12V supply voltage down to just over 5V to drive the LED. Resistor R3 is a current-limiting resistor to protect the LED.

TESTING AND CALIBRATION

Check the manufacturer's specifications to determine the threshold current of the diode you are using. For the LD60, the rated threshold current is 3 A.

Begin the calibration / test procedure by adjusting potentiometer R5 to its minimum setting.

With the 1N4005 installed in place of the laser diode, put a temporary short across the switch terminals to apply continuous power to the circuit. *Do not, under any circumstances, do this with the laser diode in the circuit. This procedure can be safely done only with a dummy diode!*

Place a voltmeter across R8 to monitor the voltage. Use Ohm's Law to determine the current:

$$I = \frac{E}{R}$$

$$I = \frac{E}{27}$$

Slowly and carefully adjust the setting of potentiometer R5 while watching the voltmeter. Adjust the potentiometer a current value just above the specified threshold. In the case of the LD60, the desired readout works out to:

$$
\begin{aligned}
E &= IR \\
&= 3 \times 30 \\
&= 90V
\end{aligned}
$$

No, you are not getting something for nothing here. The 12V supply is amplified to reach this higher value. The price for that amplification is heavy current drain, which is why heavy-duty batteries or a good ac power supply must be used with this project.

When you have found the correct setting, remove the short across switch S1. With power disconnected, remove the dummy diode (1N4005) and install the laser diode. Mount the laser diode on a small metal plate, which serves as both a support and a heat sink. Cut the perforated circuit board and metal mount for the diode to fit snugly inside the focusing tube described in Chapter 10. Construction details are illustrated in Figs. 12-7 and 12-8.

Do not focus the laser now. You might want to remove the half of the focusing tube with the lens temporarily. Aim the laser at the infrared detector described in Chapter 11 at a distance of a few inches. With the voltmeter still connected across R8, depress switch S1 briefly to take precise aim and to get an indication from the infrared detector. (The audible indicator will be a big help here, because you

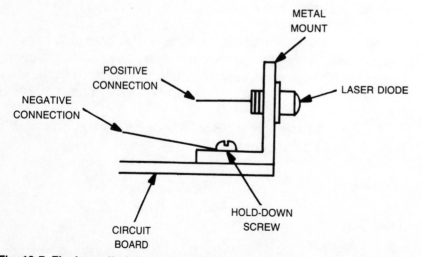

Fig. 12-7. The laser diode is mounted on a metal plate that serves as a support and as a heat sink.

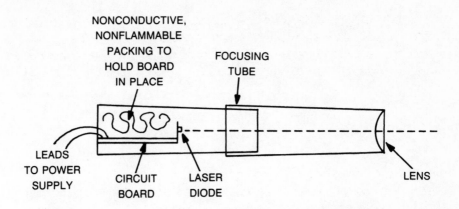

Fig. 12-8. The circuit board fits snugly inside the focusing tube.

should keep your eyes on the voltmeter as much as possible.) Make any necessary adjustments to R5 (move it very slightly). Do not hold the firing button closed for more than 1 second at any given time. Fire a brief pulse, make any needed adjustments, then fire another brief pulse.

When everything is adjusted correctly and working properly, reassemble the focusing tube with the lens. You now have a working continuous wave laser. *Remember to use all safety precautions whenever experimenting with this or any other laser project.*

Make a copy of the warning label shown in Fig. 12-1, and attach it to the laser housing.

Chapter 13

Modulated Laser Beam Transmitter

AMONG THE MANY PRACTICAL LASER APPLICATIONS ARE THOSE IN COM-
munications. There is something particularly intriguing about sending
a voice or other signal over a beam of light.

In this chapter, you will build a simple, functional laser beam
transmitter for short-range (line-of-sight) communications. The receiver
is covered in Chapter 14.

Once again, use an LD60 (or similar) laser diode that produces
an invisible infrared beam. Be sure to follow all of the precautions described in the preceding chapters.

The schematic for the transmitter is shown in Fig. 13-1, and the
parts list is in Table 13-1. Notice that a plug and jack are shown between the preamplifier and the main transmitter sections. The plug
and jack are optional, but if your budget is limited, it can be helpful
to separate the circuit into two modules. Either of the modules can
be separated easily for use in another project.

In some cases, an impedance-matching transformer might be
needed in front of the transmitter section for maximum signal clarity.
The preamplifier and the transmitter also are separated so the project can be built in two sections. The transmitter is mounted on a piece

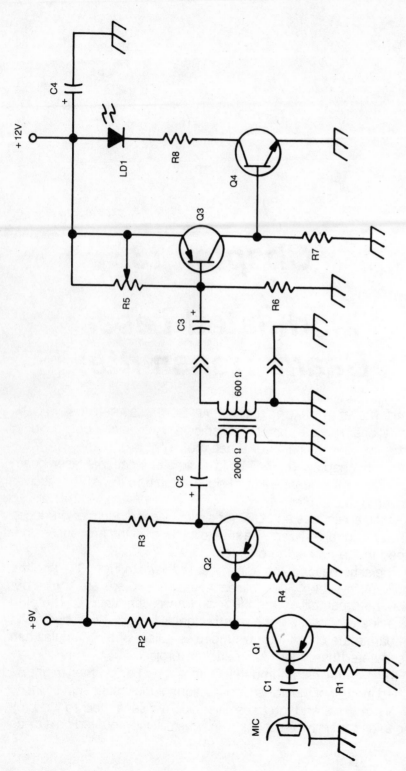

Fig. 13-1. Modulated, infrared laser transmitter.

**Table 13-1. Parts List for the Modulated
Laser Beam Transmitter of Fig. 13-1.**

COMPONENT NUMBER	COMPONENT DESCRIPTION
Q1, Q2	pnp transistor (Radio Shack RS2007, or similar)
Q3	pnp transistor (Radio Shack RS2023, or similar)
Q4	npn power tab transistor (Radio Shack RS2020, or similar)
LD1	Laser diode
C1	0.047 μF capacitor
C2, C3	2.5 μF, 16 V electrolytic capacitor
C4	100 μF, 25 V electrolytic capacitor
R1, R4	4.7 kΩ ¼ W resistor
R2, R3, R7	1 kΩ ¼ W resistor
R5	2.5 kΩ trimmer potentiometer
R6	47 kΩ ¼ W resistor
R8	270 Ω ½ W resistor
T1	impedance-matching transformer (2 kΩ: 600 Ω)

of perforated circuit board, which is placed inside the focusing tube. The preamplifier will be housed in a separate minibox.

The audio signal is picked up by the microphone. You can use a replacement microphone for a cassette tape recorder for most applications. Remember, the signal fidelity at the receiver can be no better than the signal produced by the microphone. If you want a high-quality signal at the receiver, you must use a high-quality microphone. If high fidelity is not that important to you, use a less expensive microphone. For general experimentation and standard voice communications, you should be able to buy a suitable microphone for about five dollars.

As mentioned previously, the project is constructed in two sections. Figure 13-2 illustrates the components mounted on the preamplifier board. The components on the transmitter board are shown in Fig. 13-3. Cut the transmitter board to fit snugly inside the focusing tube as shown in Fig. 13-4.

Be sure to use adequate heatsinking for the laser diode. The heatsinking is not quite as critical as with the continuous-wave laser

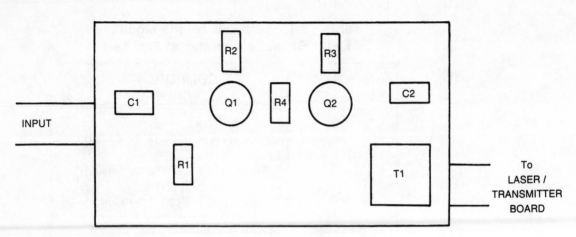

Fig. 13-2. Half of the circuit is mounted on the preamplifier board.

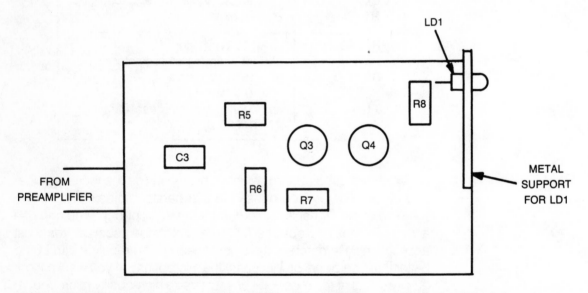

Fig. 13-3. The main laser and transmitter circuits are mounted on a second circuit board.

of Chapter 12, because in this case the laser is being powered intermittently by the audio signal. However, overheating can still be a problem, as with any laser device. It is always better to use too much heatsinking rather than too little.

Once you have completed construction, carefully double-check all of the wiring and solder joints before applying power. (This should be done for any project, of course.)

Until you build a receiver, all you can do is check to make sure that the laser is putting out a signal. Test this with the infrared detector

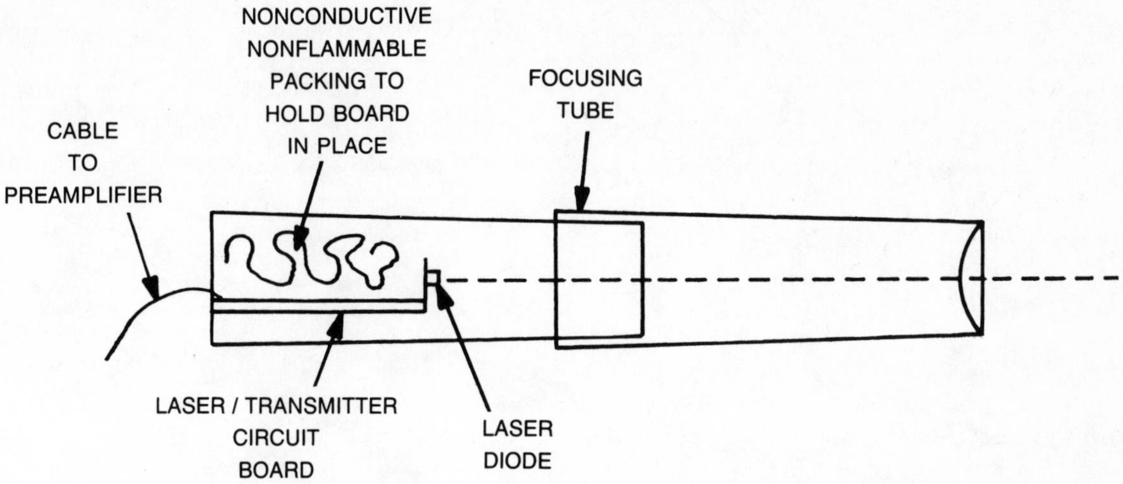

Fig. 13-4. The laser/transmitter board is cut to fit snugly inside the focusing tube.

project of Chapter 11. Feed some signal into the microphone as you are testing it. By adjusting the sensitivity control on the detector, you can also determine the difference in response for loud and soft sounds picked up by the microphone.

Without a receiver, there is not much else you can do with this project, so move on to the next project in Chapter 14: the modulated laser beam receiver.

Chapter 14

Modulated
Laser Beam Receiver

IN THIS CHAPTER YOU WILL CONSTRUCT THE COMPANION RECEIVER FOR
the modulated laser beam transmitter of Chapter 13.

THE RECEIVER CIRCUIT

The schematic for this project is shown in Fig. 14-1, and the parts
list is in Table 14-1.

The modulated laser beam from the transmitter is detected by
the phototransistor (Q1). The electrical signal formed across this
component is fed through a preamplifier and out through a speaker
or headphones. In most cases use an impedance-matching
transformer to match the circuit output to the speaker used. This circuit
output impedance is about 1,000 Ω. Most speakers have a nominal
impedance of 8 Ω.

In use, the transmitter and receiver are positioned so that the
transmitter beam is focused directly on the receiver sensor
(phototransistor). Because the infrared laser beam is invisible, use a
visible light source, like the laser simulator of Chapter 10, for initial
positioning. Then remove the simulator and position the transmitter
as closely as possible in the same position. Feed a continuous signal
through the transmitter (a steady tone would be best, but you can

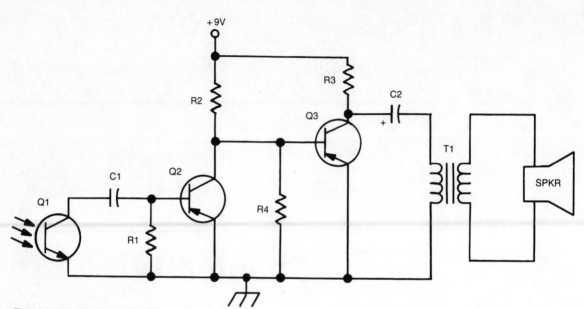

Fig. 14-1. Laser beam receiver receives and demodulates the signal from the transmitter circuit of Chapter 13.

Table 14-1. Parts List for the Modulated Laser Beam Receiver of Fig. 14-1.

COMPONENT NUMBER	COMPONENT DESCRIPTION
Q1	Phototransistor (infrared sensitive)
Q2, Q3	pnp transistor
	(Radio Shack RS2007 or similar)
C1	0.047 µF capacitor
C2	2.5 µF 16 V electrolytic capacitor
R1, R4	4.7 kΩ resistor
R2, R3	1 kΩ resistor
T1	impedance-matching transformer
	(2 kΩ: 8 Ω)*
SPKR	speaker (or headphones)
* Select secondary to match speaker or headphones.	

just tap the microphone). Listen to the receiver for the strongest signal as you carefully move the transmitter in small increments.

For initial testing, position the transmitter and receiver within a few inches of each other. For practical applications, you will obviously want more distance between the transmitter and the receiver. Setup and focusing is easier if you have someone to help you.

LIMITATIONS

You will probably be pleasantly surprised by the quality of the transmitted signal from these projects. But there are certain definite limitations.

In some environments, you might pick up interference from other light sources. You might be able to shield the receiver sensor from some of the ambient light. If possible, reduce the ambient lighting. Often performance can be significantly improved by placing a red filter over the phototransistor, allowing it to ignore much of the interfering light. Of course, a simple filter does little good if the interfering light is in the deep red or infrared region of the spectrum.

Another major limitation of this transmitter-receiver combination is that it is strictly a line-of-sight system. It cannot operate around corners. Any object between the transmitter and the receiver will block the laser beam signal.

If the transmitter and receiver are any distance apart, there is a real chance of someone moving between them and possibly looking into the invisible laser beam without realizing it. Avoid this situation at all costs.

This is an experimental system with limited practical applications unless it is modified for a specific purpose. Several modifications are described in the following sections.

FIBEROPTICS

One good solution to the limitations of the modulated laser beam communications system is to use a fiberoptic cable to carry the modulated light beam from the transmitter to the receiver. A fiberoptic cable is made of glass or plastic fibers; each fiber is thinner than a human hair. Optic fibers can conduct light—even around sharp corners—with extremely low losses. A fiberoptic cable is a special wire that carries light beams rather than electrical current.

Ordinarily, light waves (including laser beams) can travel only in straight lines. If anything gets in the way, the light is stopped. The light cannot go around any obstacle. With a fiberoptic cable, light waves can be forced to follow any curve—even a sharp angle—thanks to refraction. The refractive index of the fiber core is slightly higher than that of its *cladding* (outer surface). This effectively confines the light waves to the interior of the fiber, regardless of any curves or bends, as illustrated in Fig. 14-2.

One end of the fiberoptic cable is connected to the light source (laser diode), and the other end is connected to the light-sensitive re-

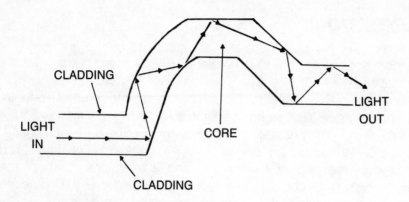

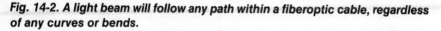

Fig. 14-2. A light beam will follow any path within a fiberoptic cable, regardless of any curves or bends.

ceiver (the phototransistor). Special fiberoptic connectors are available for this purpose.

Using a fiberoptic cable will correct most of the system limitations discussed earlier, but the cost is that the convenience of wireless communication is lost. You can determine which option is better for your application.

OTHER APPLICATIONS

This receiver circuit can be used to listen to any varying light source. (A different phototransistor might be needed to maximize the sensitivity at the desired light frequency.) In a darkened room, aim the sensor at a television screen. You will hear some interesting sounds fluctuating in step with changes in the picture. You could say this is a color organ in reverse. Another interesting experiment is to use the light receiver to monitor lightning flashes on a stormy night. No, you do not have to stand out in the rain; you can aim the sensor through a window.

Use your imagination to come up with your own applications and variations for all the projects in this book.

Chapter 15

Advanced Technical Data

THE PROJECTS IN THIS BOOK (CHAPTERS 10 THROUGH 14) ARE FAIRLY SIM-
ple, experimental devices. Their purpose is to get you started working with lasers. You are encouraged to spread out and start
experimenting further on your own, using these projects as starting
points. This chapter offers additional information you might find useful
in your future laser experiments.

MIXED CRYSTALS

One of the newest lasing materials is the mixed crystal GaAsSb,
introduced by RCA. It is a mixture of two crystals: gallium arsenide
and gallium antimonide. Mixed crystals in general have a special property in that the wavelength of the light they emit can be tuned by
adjusting their composition. For example, by increasing the amount
of antimony in GaAsSb, the light emission can be continuously tuned
to any wavelength from 0.9 to 1.2 micrometers.

Mixed crystals are not new inventions. Gallium arsenide phosphide
has been used in commercial red LEDs. Aluminum gallium arsenide
is an integral part of the GaAs lasers now being used in the Atlanta
Lightwave Communication System. However, do not assume that
mixed crystals can be formed for any given set of elements. Whether

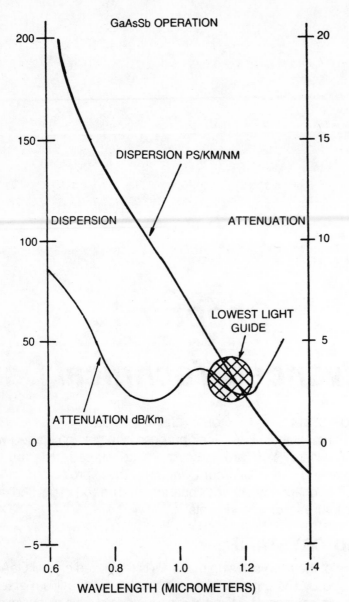

Fig. 15-1. Minimum-loss chart exhibiting dispersion and attenuation.

or not a solid mixed crystal of a given composition can be formed depends on the forces between the particular atoms in the crystal and on the applicable laws of thermodynamics. For GaAsSb, the limit for the GaSb content is more than 25%. This makes it possible to have lasers with wavelengths as long as 1.2 micrometers.

The new laser consists of a very thin layer of GaAsSb sandwiched between P-type and N-type AlGaAs wafers. These two wafers or layers inject electrons and holes into the GaAsSb active layer and provide barriers that confine the electrons and holes in the GaAsSb layer. The confinement of both electric charges and light is a key feature of this layered *double heterostructure.* This makes possible the efficient operation of the laser.

The multiple layers for the lasers are grown from liquid solutions of the elements by epitaxial techniques. With this technique, as a layer solidifies, it assumes the crystal structure of the layer beneath it. The process is similar to that used for GaAs and is only slightly more complicated because of the addition of antimony. It is fabricated by first growing a wafer of about 1 square centimeter. The opposite sides are then silvered. Following that, they are diced to yield devices from approximately 0.010 to 0.015 inches.

Electrically, the GaAsSb laser exhibits exceptional dispersion and attenuation characteristics (Fig. 15-1).

EXCITATION OF MOLECULES

A certain amount of energy is required to achieve excitation. The energy is expressed in terms of calories per mole. A mole represents a standard number of molecules.

Electronic excitation requires the highest input of energy, equivalent to that of ultraviolet or visible light. The other forms of excitation require less energy (Fig. 15-2).

Chemical Pumping

Chemical pumping is based on the energy released in the making and breaking of bonds. In Fig. 15-3A, atom A might combine with a molecule consisting of atoms B and C. This would produce an intermediate and transient molecule possessing extra energy. This molecule could separate into two molecular fragments (Fig. 15-3B). Either one might be excited and could be stimulated enough to drop to a lower energy level (Fig. 15-3C), emitting a photon in the process.

Hydrogen and Chlorine Reaction

Reactions between hydrogen and chlorine in an explosion provide pumping for a chemical laser. A trigger of light (Fig. 15-4A) separates a chlorine molecule into two chlorine atoms. One of them (Fig. 15-4B) combines with a hydrogen molecule to yield a hydrogen chlorine molecule and a free hydrogen atom. This produces an excited hydrogen chlorine molecule (Fig. 15-4C) that emits a photon of infrared radiation (Fig. 15-4D).

TYPE	EXCITATION	ENERGY LEVEL	MECHANISM
ELEC-TRONIC	VISIBLE OR ULTRAVIOLET LIGHT	100,000 TO 200,000 CALORIES PER MOLE	
VIBRA-TIONAL	INFRARED LIGHT	500 TO 10,000 CALORIES PER MOLE	
ROTA-TIONAL	MICROWAVE LIGHT	0.1 TO 100 CALORIES PER MOLE	
TRANS-LATIONAL	HEAT	ANYTHING ABOVE ZERO	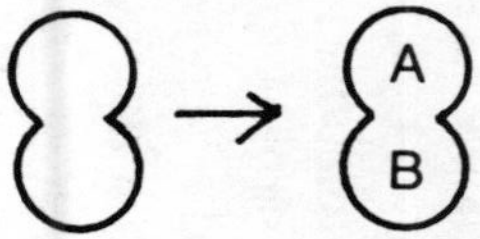

Fig. 15-2. Electronic excitation requires different levels of energy, depending on the type of excitation.

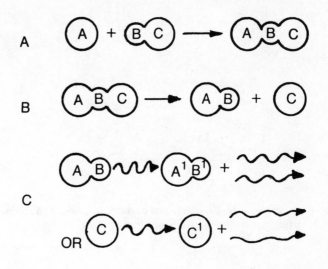

Fig. 15-3. *The process of chemical pumping makes and breaks bonds.*

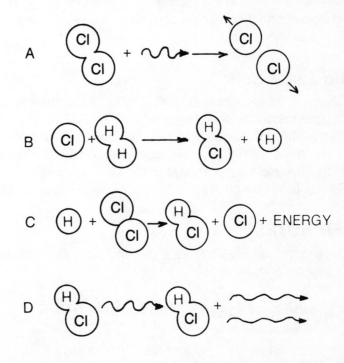

Fig. 15-4. *When hydrogen and chlorine react, they produce an excited hydrogen chloride molecule that emits an infrared photon.*

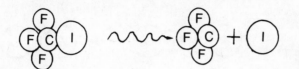

Fig. 15-5. In an iodine laser, a molecule of carbon, fluorine, and iodine releases an excited iodine atom that emits a photon of light energy.

Iodine Laser

The iodine laser derives its pumping from the dissociation by light of a molecule consisting of a carbon atom, three fluorine atoms, and an iodine atom (CF_3I). In the rupture of the carbon-iodine bond, an excited iodine atom (Fig. 15-5) is born. This releases a photon of light energy, as shown in Fig. 15-5B.

Liquid Laser Cells

Typical liquid laser cells perform the function of the resonating cavities in solid and gaseous lasers (Fig. 15-6). All liquid lasers have expansion receivers to accommodate the expansion and contraction of the liquid during thermal changes that occur during normal operation. The lasers are optically pumped by means of surrounding flash lamps. These lamps are of either the spiral or the proximity type.

LASER NOTES

This section features some equations and definitions that are useful to you as a laser experimenter.

Metric Units of Linear Measurement

1 micron = 1 millionth of a meter (10^{-6} meter)
1 millimicron = 1 millionth of a millimeter (10^{-6} millimeter)
1 angstrom = 10^{-8} centimeter (10^{-10} meter)
1 nanometer = 10^{-7} centimeter (10^{-9} meter)

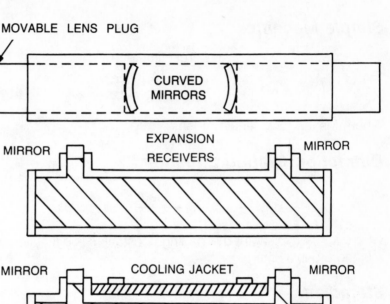

MOVABLE LENS PLUG

CURVED MIRRORS

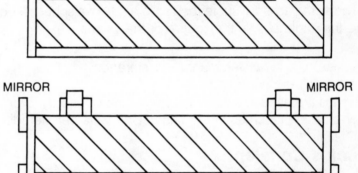

MIRROR ··· EXPANSION RECEIVERS ··· MIRROR

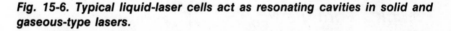

MIRROR ··· COOLING JACKET ··· MIRROR

MIRROR ··· MIRROR

Fig. 15-6. Typical liquid-laser cells act as resonating cavities in solid and gaseous-type lasers.

Equivalents of Vacuum Measurements

1 atmosphere = 29.92 inches of Hg (mercury)
= 760 millimeters of Hg
1 millimeter of Hg = 0.019 psi (pounds per square inch)
= 0.0446 foot of water
1 inch of water = 1.87 millimeters of Hg
= 0.036 psi
1 inch of Hg = 25.4 millimeters of Hg
= 13.6 inches of water
1 millimeter of Hg = 1 torr
= 1/760 of an atmosphere

Simple Magnifier

$$M = \frac{25 \text{ centimeters}}{f}$$

(approximately)

Where f is the focal length of lens.

Diffraction Grating Equation

$$\lambda = \frac{d \text{ sine } \theta}{n}$$

Where λ = wavelength; d is grating constant; θ is diffraction angle; n is the order of image.

Steradian

The steradian is the subtended solid angle of a spherical portion, where the angle equals the square of the radius (r^2).

$$\pi \text{ steradians } = 1 \text{ steregon}$$

Steradians are used in measures of brilliance or energy to describe power of lasers (such as in the number of watts per steradian).

Planck's Constant

$$6.62 \times 10^{-34} \text{ joules per second}$$

Planck's Formula

$$E = hf$$

Where E = energy of a photon in joules, f = vibrations per second, (hertz) of radiation, and h = Planck's constant: 6.62×10^{-34} joules per second (6.6×10^{27} erg seconds)

TYPE OF RADIATION	MEAN VALUE OF ϕ (Hz)	hf IN JOULES
Radio waves	3×10^5	2.0×10^{-28}
Heat waves	3×10^{13}	2.0×10^{-26}
Visible light	6×10^{14}	3.9×10^{-19}

TYPE OF RADIATION	MEAN VALUE OF ϕ (Hz)	hf IN JOULES
X-rays	3×10^{18}	2.0×10^{-15}
Gamma rays	3×10^{19}	2.0×10^{-14}

Mass of a Photon

$$m = \frac{hf}{c^2}$$

Where m = the photon mass; h = Planck's constant, f = the frequency in Hz, and c = the speed of light.

Energy of a Photon

$$E = h \times f$$

Where E = the energy in joules, h = Planck's constant, and f = the frequency in hertz.

Wavelength of a Photon

$$\lambda = \frac{h}{mc}$$

Where λ = photon wavelength; h = Planck's constant; m = mass of photon; c = the velocity of light (mc is photon momemtum).

Wavelength of a Particle Having a Velocity

$$\lambda = \frac{h}{mv}$$

Where λ = the particle wavelength, h is Planck's constant, m = mass of particle, v = velocity (mv is particle momentum).

Einstein's Photoelectric Equation

$$\tfrac{1}{2}mv^2max = hf - w$$

Where f = vibrations per second (Hz), h = Planck's constant 6.62 × 10^{-34}, and w = the work function of the substance.

Speed of Light in Vacuum

300,000,000 meters per second (3 × 10^8 meters per second) or 186,000 miles per second

one kilowatt hour = 3,600,000 joules

1 gram mass converted to energy = 9 × 10^{20} ergs = 9 × 10^{13} joules

Glossary

absolute zero—The lowest possible temperature: $-273.16°C$ or $-459.69°F$

absorption spectrum—A continuous spectrum, like white light, interrupted by dark lines or bands that are produced by the absorption of certain wavelengths by a substance through which the light or other radiation passes.

ampere—The mks (meter kilogram second) unit for measuring electric current (it equals the flow of one coulomb per second).

amplifier—An electronic device that changes a weak signal into a stronger one.

amplitude—The maximum displacement, or graphic height, of an oscillating wave. It is measured from time line to peak top.

angstrom (A)—A unit of length equal to 10^{-10} of a meter, used in measuring the wavelength of light. Atoms have a radius of from one to two angstroms.

atom—The smallest particle of an element that has all of its chemical properties; it is composed of at least one proton, one neutron, and one electron.

atomic number—The number of protons in the nucleus of an atom.

beam divergence in a given plane—The half of divergence of the laser emission at which the intensity of radiation is one half the peak intensity.

beta rays—Streams of fast moving particles (electrons) ejected from radioactive nuclei.

cathode rays—Electrons emitted by a cathode, as in an electron gun.

cgs (centimeter gram second)—The metric system of measurement in which the fundamental units ae the centimeter, the gram, and the second.

chromatic aberration—The inability of a single lens to refract all the different colors of light to the same focus.

coherent radiation—Radiation in which the phase difference between any two points in the radiation field is constant throughout the duration of the radiation.

color—The property of light that depends on its frequency. It is visible to the eye.

component—Often referring to one of the several vectors combined algebraically or geometrically to yield a resultant vector.

concave lens—A lens that diverges parallel light rays to a focal point.

conservation of matter and energy—A law that expresses that the total amount of energy and matter in the universe is constant, which can be equated as $E = mc^2$.

constructive interference—The superposition of two waves approximately in phase so that their amplitudes add up to produce a combined wave of larger amplitude than the components.

continuous spectrum—A spectrum consisting of a wide range of unseparated wavelengths.

converging lens—A lens that is thicker in the middle than it is at the edge.

convex lens—A lens that converges parallel light rays.

cosmic rays—High energy particles, apparently from beyond our solar system.

coulomb—The quantity of electricty equal to the charge on 6.25×10^{18} of electrons.

De Broglie matter waves—Associated wave properties of particles of matter. The wavelength of a particle is related to its momentum and Planck's constant *h* by the relationship:

$$\text{Wavelength} = \frac{h}{\text{momentum}}$$

decay—The act of an atom or population of atoms falling from an excited level of energy. Decay can be thought of as the reverse of stimulation or excitation.

dichroism—A property of certain crystalline substances in which one polarized component of incident light is absorbed and the other is transmitted.

diffraction grating—An optical surface, either transmitting or reflecting, with several thousand equally spaced and parallel grooves ruled in it.

dispersion—The separation of polychromatic light into its component wavelengths.

diverging lens—A lens that is thicker at the edge than it is in the middle.

duty factor (du)—The product of the pulse duration and the pulse repetition frequency of a wave composed of pulses that occur at regular intervals.

electromagnetic waves—Transverse waves in space, having an electric component and a magnetic component, each being perpendicular to the other and both being perpendicular to the direction of propagation.

electron—A negatively charged atomic particle having a rest mass of 9.1083 × 10^{28} grams.

electron shell—A region about the nucleus of an atom in which electrons move or orbit.

elementary colors—The six regions of color in the solar spectrum, observed by the dispersion of sunlight (red, orange, yellow, green, blue, and violet).

emission—The release of radiant energy. The act of discharging excess energy.

energy level—A region about the nucleus of an atom, where electrons are orbiting, also representing the particular energy status of the electron.

erg—The cgs system unit of work, a force of one dyne acting through a distance of one centimeter.

excitation—The process of boosting one or more electrons in an atom or molecule from a lower to a higher energy level. An atom in such an excited state will usually decay rapidly to a lower state or level accompanied by the emission of radiation. The frequency and energy of emitted radiation are related by: $E = hf$.

F-number—The ratio of the focal length of a lens to the effective aperture.

focal length—The distance between the principal focus of a lens and its optical center or vertex.

focus—A point at which light rays meet or from which rays of light diverge.

focus, principal—A point to which rays parallel with the principal axis converge, or from which they diverge, after reflection or refraction.

Fraunhofer lines—Absorption lines in the solar spectrum.

frequency—Number of vibrations or cycles per second (cps or Hertz).

frequency, cutoff—A characteristic threshold frequency of incident light, below which, for a given material, the photoelectric emission of electrons ceases.

gamma ray—A high-energy wave emitted from the nucleus of a radioactive atom.

ground level—The energy level of the atomic population that occurs naturally or in absence of additive stimulation. The relative level before population inversion is effected, from which the level of excitation after inversion can be measured. It is the base line or reference point in considerting the degree of excitation.

index of refraction—The ratio of the speed of light in a vacuum to its speed in any other given substance.

infrared-emitting diode—A semiconductor device in which focused, injected minority carriers produce infrared light radiation flux when current flows as a result of applied voltage.

injection laser—A solid-state semiconductor device consisting of at least one PN junction capable of emitting coherent or stimulated radiation under specified conditions. The device incorporates a resonant optical cavity.

interference—The superimposition of one wave on another, in either a constructive or destructive manner. The mutual effect of two beams of light, resulting in a loss of energy in certain areas and reinforcement of energy in others.

joule—The mks unit of work, a force of one newton acting through a distance of one meter (Nt • m).

kilowatt hour—A unit of electric energy equal to 3.6×10^6 W per second.

laser—A device that emits coherent monochromatic light by the process of light ampliication by the stimulated emission of radiation. An optic maser: a device that emits coherent, amplified radiations on the subvisible range. Being in the microwave category, accomplished by microwave amplification by the stimulated emission of radiation, the maser was invented first and the laser was a ramification of it.

lasing condition (or state)—The condition of an injection laser corresponding to the emission of predominantly coherent or stimulated radiation.

mass—The measure of the quantity of matter.

matter—Anything that occupies space and has mass.

meter—The basic unit of length in the metric system (39.37 inches).

mole—That quantity of a substance whose mass in grams is numerically equal to the mass of one of its molecules in atomic mass units.

molecule—The smallest particle that an element or compound can exist as and be chemically independent.

nanometer (nm)—Unit of measurement sometimes used in light technology. It is equal to 10^{-9} meters or 3.937×10^{-8} inches.

neutron—A neutral atomic particle having a mass of 1.675×10^{-24} grams.

nucleus—The positively charged dense central part of an atom.

optical center—The point in a lens through which the secondary axis passes.

optical density—A property of a transparent material that is a measure of the speed of light through it.

paramagnetism—The property of a substance by which it is attracted to a strong magnet.

phase—The position and motion of a particle of a wave.

photoelectric effect—The emission of electrons by a substance when illuminated by electromagnetic radiation.

photoelectric effect, inverse—The emission of photons of radiation due to the bombardment of a material with high speed electrons.

photoelectric emission, 1st. law of—The law stating that the rate of emission of photoelectrons is directly proportional to the intensity of the incident light.

photoelectric emission, 2nd. law of—The law stating that the kinetic energy of photoelectrons is independent of the intensity of the incident light.

photoelectric emission, 3rd. law of—The law stating that within the region of effective frequencies, the maximum kinetic energy of photoelectrons is directly proportional to the frequency of the incident light.

photoelectrons—Electrons emitted from a light sensitive material when it is properly illuminated.

photon—A quantum of light energy whereby Planck's constant h relates the energy and frequency of light quanta: $h = 6.6 \times 10^{-34}$ joules per second. Photons are also thought of as being in the form of packets of energy.

piezoelectric effect—The property of certain natural and synthetic crystals to develop a potential difference (voltage) between opposite surfaces when subjected to mechanical stress or vibrations.

Planck's constant—A universally proportional constant, relating photon energy to the frequency of radiation ($h = 6.6 \times 10^{-34}$ joules per second).

plasma—Hot gases composed of electrically charged particles. Most of the matter in the universe is in the plasma state.

population—The total of the atoms in a mass.

population inversion—The process of inverting the population of atoms in a mass from one state to another. To raise the normally occurring small percentage of the atomic population which exists in an excited state, to a greater majority or percentage of this population to a higher level of excitation, thus inverting the percentage or ratio of the population from a lower to a higher level.

population reversion—The decay of the population to the energy level that existed before the population was first inverted, accompanied by the release of the previously applied energy. It is emitted in the form of a packet of energy, also known as a photon.

prelasing condition (or state)—The condition of an injection laser corresponding to the emission of predominantly incoherent or spontaneous radiation.

primary colors—Red, blue, and green. From these, all other secondary colors can be achieved.

principal axis—A line drawn through the center of curvature and the optical center of a lens.

proton—A positively charged atomic particle having a mass of 1.67×10^{-4} grams.

pulse—A singularly nonrecurring disturbance, such as a laser pulse.

quantum efficiency (QE)—The ratio of the quanta of radiant energy (photons) emitted per second to the number of electrons flowing per second, for example, photons / electron.

quantum theory—The theory that the transfer of energy between light radiations and matter occurs in discrete units or packets, the magnitude depending on the frequency of the radiation.

radiant efficiency of a source of radiant flux (η)— The ratio of the total radiant flux to the forward power dissipation.

radiant flux (radiant power) (Φ)—The time rate of flow of radiant energy. It is expressed in watts (preferably) or ergs / second.

Radiant intensity (I)—The radiant flux proceeding from the source per unit solid angle in the direction considered, watts / steradian.

radiation—The form of energy in transit that occurs when an atom decays, or reverts to ground level, after radiating the necessary energy for it to do so. Radiation can, and quite often does, manifest itself as visible radiation or light in the form of photons, or discrete packets of energy.

radiation pattern—The representation of the intensity of emission as a function of direction in a given plane. The axes are to be specified with respect to the junction plane and the cavity face.

rectilinear propagation—Energy waves travelling in a straight line.

refraction—The bending of a wave disturbance as it passes obliquely from one medium into another of different density.

resonance—1. The inducting of vibrations or oscillations of a natural rate in matter by a vibrating or oscillating source, having the same or a multiple related frequency. 2. The condition whereby energy waves can be amplified by the continued production of waves having a duplicated length as the cavity length where such waves are produced. Such resonant cavities are utilized in the maser and laser.

resultant—Of or referring to a vector representing the algebraic or geometric sum of several components.

rise time (t_r)—The time taken for the radiation flux to increase from 10% to 90% of its peak value when the laser is subjected to a step function current pulse of specified amplitude.

scattering—Emission of light in every direction, caused by the time-varying electric vector of incident light setting electrons into oscillation.

secondary axis—Any line, other than the principal axis, drawn through the center of curvature of the optical center of a lens.

secondary emission—Emission of electrons as a result of the bombardment of an electrode by high-velocity electrons.

spectral bandwidth ($\Delta\lambda$)—The spectral bandwidth for single peak devices is the difference between the wavelengths at which the radiant intensity is 50% (unless otherwise stated) of the maximum value.

spectral radiant flux ($\Phi\lambda$)—The radiant flux per unit wavelength interval at wavelength λ, watts/nanometer.

stimulation—An event or force that causes an atom to increase or decrease its value in some respect or triggers a reaction in an atom.

superposition—The process of combining the displacements of two or more wave motions, algebraically to produce a resultant wave motion.

thermionic emission—The liberation of electrons from the surface of a heated body, usually occurring at a point of visible incandescence.

threshold current (I_{th})—The minimum forward current for which the laser is in a lasing state at a specified temperature.

threshold frequency—The minimum frequency of incident light that will eject a photon from a given metal or substance.

transverse wave—A wave in which the particles of the medium vibrate at right angles to the path along which the wave travels through the medium.

Vander Waals' forces—Attractive forces arising from the effect of the varying electric field of atoms of one molecule on the electric field of atoms of another molecule.

vector quantity—A quantity which requires both a magnitude and a direction for its complete description.

watt—The mks system unit of power (one joule per second, 1V × 1A).

wavelength—The distance from one particle or peak to the next following particle or peak measured along a parallel line with the time line.

wavelength of peak radiant intensity—The wavelength at which the spectral distribution of radiant intensity is a maximum.

X-rays—Electromagnetic radiations of very short wavelengths and high frequency, enabling great penetrating power into and through substances.

Appendix A

The Periodic Table

The modern periodic table (Fig. A-1), which differs only slightly from the table compiled by Dimitri Mendeleev in 1869, is a tabular description of all of the naturally occurring elements (92) and the additional 14 synthesized, man-made elements. These are a total of 106 occupations in the periodic system.

Reference is made to the periodic table and to Figs. A-2 and A-3. For purposes of explanation, iron (Fe) has been selected as an example of what the periodic table tells about each element of the system.

In the upper left hand corner (Fig. A-2) is a number ranging from 1 to 103. This number is the atomic number of the element. It tells us two things about the element: the number of protons in the nucleus and the number of electrons that the atom will normally have in a neutral state. There will be present one electron in the outer orbital envelope for each proton present in the nucleus.

In the center of the box is the *chemical symbol* or abbreviation for the element. In this case iron is chemically symbolized as *Fe*.

Below the chemical symbol is the atomic weight of the atom. This weight represents the weight of the element. It is a measure of the mass of the atom and is the number of protons and neutrons. Thus,

Fig. A-1. The periodic table.

PARENTHETICAL VALUES ARE MASS NUMBERS OF THE ISOTOPES WITH LONGEST HALF LIVES

TRANSITION ELEMENTS

PERIOD COLUMN	I	II												III	IV	V	VI	VII	0	ORBITALS BEING FILLED
n = 1	1 H 1.00																		2 He 4.00	1s
n = 2	3 Li 6.94	4 Be 9.01												5 B 10.8	6 C 12.01	7 N 14.01	8 O 16.00	9 F 19.0	10 Ne 20.2	2s2p
n = 3	11 Na 23.0	12 Mg 243.												13 Al 27.0	14 Si 28.1	15 P 31.0	16 S 32.1	17 Cl 35.5	18 Ar 39.9	3s3p
n = 4	19 K 39.1	20 Ca 40.1	21 Sc 45.0	22 Ti 47.9	23 V 50.9	24 Cr 52.0	25 Mn 54.9	26 Fe 55.8	27 Co 58.9	28 Ni 58.7	29 Cu 63.5	30 Zn 65.4		31 Ga 69.7	32 Ge 72.6	33 As 74.9	34 Se 79.0	35 Br 79.9	36 Kr 83.8	4s3d4p
n = 5	37 Rb 85.5	38 Sr 87.6	39 Y 88.9	40 Zr 91.2	41 Nb 92.9	42 Mo 95.9	43 Tc (99)	44 Ru 101.1	45 Rh 102.9	46 Pd 106.4	47 Ag 107.9	48 Cd 112.4		49 In 114.8	50 Sn 118.7	51 Sb 121.8	52 Te 127.6	53 I 126.9	54 Xe 131.3	5s4d5p
n = 6	55 Cs 132.9	56 Ba 137.3	57-71 See Below	72 Hf 178.5	73 Ta 180.9	74 W 183.9	75 Re 186.2	76 Os 190.2	77 Ir 192.2	78 Pt 195.1	79 Au 197.0	80 Hg 200.6		81 Tl 204.4	82 Pb 207.2	83 Bi 209.0	84 Po (209)	85 At (210)	86 Rn (222)	6s4f5d6p
n = 7	87 Fr (223)	88 Ra (226)	89- See Below																	7s5f6d7p

PERIOD																ORBITALS BEING FILLED
n = 6	57 La 138.9	58 Ce 140.1	59 Pr 140.9	60 Nd 144.2	61 Pm (147)	62 Sm 150.4	63 Eu 152.0	64 Gd 157.3	65 Tb 158.9	66 Dy 162.5	67 Ho 164.9	68 Er 167.3	69 Tm 168.9	70 Yb 173.0	71 Lu 175.0	4f
n = 7	89 Ac (227)	90 Th (232)	91 Pa (231)	92 U 238.0	93 Np (237)	94 Pu (242)	95 Am (243)	96 Cm (247)	97 Bk (249)	98 Cf (251)	99 Es (254)	100 Fm (253)	101 MD (256)	102 No (253)	103 Lw (257)	5f

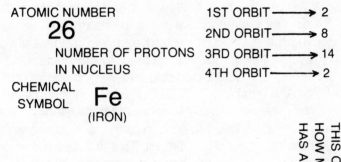

ATOMIC NUMBER
26
NUMBER OF PROTONS
IN NUCLEUS
CHEMICAL
SYMBOL **Fe**
(IRON)

1ST ORBIT ⟶ 2
2ND ORBIT ⟶ 8
3RD ORBIT ⟶ 14
4TH ORBIT ⟶ 2

THIS COLUMN INDICATES
HOW MANY ORBITS THE ATOM
HAS AND THE NUMBER OF ELECTRONS

ATOMIC WEIGHT = NUMBER OF PROTONS
55.8
(ROUNDED OFF TO 56)

Fig. A-2. Description of iron in the periodic table.

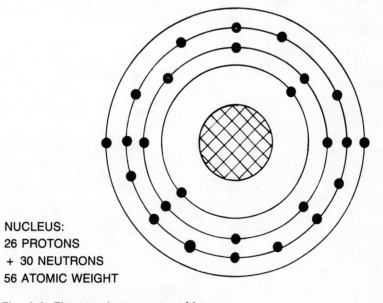

NUCLEUS:
26 PROTONS
+ 30 NEUTRONS
56 ATOMIC WEIGHT

Fig. A-3. The atomic geometry of iron.

you may very easily determine the number of neutrons of the element by rounding off the atomic weight to the nearest whole number and subtracting the number of protons as given in the atomic number. The answer is the number of neutrons in the nucleus.

Referring to the periodic table (Fig. A-1), observe on the far left-hand side, the notation *PERIOD COLUMN*. This shows the number of electron orbits the atom has in its outer envelope. Note that iron (Fe) has four such orbits. The column on the far right shows the number of electrons in each individual orbit. This enables you to construct a geometric drawing of each element from all of the notations given.

The number of electrons in the outermost orbit also tells the valence of the element. The combining ability of each atom is governed by the number of electrons, or hooks, in its outer orbit. Because iron has an outer number of two electrons, it has a valence of two. Toward the right-hand side of the table is a stair-step division of the elements. This has been added to separate the elements into metals and nonmetals. The 84 elements to the left of the division are metals, and those to the right are nonmetals. To the extreme right, under the heading 0, are the inert elements, which by virtue of their completed outer orbits, will not combine with any elements whatsoever. These elements are *chemically inert*.

The original table, as compiled by Mendeleev, had the elements arranged in successive order by their atomic weights. The modern revision arranges the progression according to the element proton count; this arrangement is more practical for scientific purposes.

Appendix B

Atomic Weights
of the Elements

ATOMIC WEIGHTS OF THE ELEMENTS							
Element	Symbol	Atomic Number	Atomic Weight	Element	Symbol	Atomic Number	Atomic Weight
Actinium	Ac	89	(227)	Mercury	Hg	80	200.59
Aluminum	Al	13	26.98	Molybdenum	Mo	42	95.94
Americium	Am	95	(243)	Neodymium	Nd	60	144.24
Antimony	Sb	51	121.75	Neon	Ne	10	20.183
Argon	Ar	18	39.948	Neptunium	Np	93	(237)
Arsenic	As	33	74.92	Nickel	Ni	28	58.71
Astatine	At	85	(210)	Niobium	Nb	41	92.91
Barium	Ba	56	137.34	Nitrogen	N	7	14.007
Berkelium	Bk	97	(249)	Nobelium	No	102	(253)
Beryllium	Be	4	9.012	Osmium	Os	76	190.2
Bismuth	Bi	83	208.98	Oxygen	O	8	15.9994
Boron	B	5	10.81	Palladium	Pd	46	106.4
Bromine	Br	35	79.909	Phosphorus	P	15	30.974
Cadmium	Cd	48	112.40	Platinum	Pt	78	195.09
Calcium	Ca	20	40.08	Plutonium	Pu	94	(242)

Element	Symbol	Atomic Number	Atomic Weight	Element	Symbol	Atomic Number	Atomic Weight
Californium	Cf	98	(251)	Polonium	Po	84	(210)
Carbon	C	6	12.011	Potassium	K	19	39.102
Cerium	Ce	58	140.12	Praseodymium	Pr	59	140.91
Cesium	Cs	55	132.91	Promethium	Pm	61	(147)
Chlorine	Cl	17	35.453	Protactinium	Pa	91	(231)
Chromium	Cr	24	52.00	Radium	Ra	88	(226)
Cobalt	Co	27	58.93	Radon	Rn	86	(222)
Copper	Cu	29	63.54	Rhenium	Re	75	186.23
Curium	Cm	96	(247)	Rhodium	Rh	45	102.91
Dysprosium	Dy	66	162.50	Rubidium	Rb	37	85.47
Einsteinium	Es	99	(254)	Ruthenium	Ru	44	101.1
Erbium	Er	68	167.26	Samarium	Sm	62	150.35
Europium	Eu	63	151.96	Scandium	Sc	21	44.96
Fermium	Fm	100	(253)	Selenium	Se	34	78.96
Fluorine	F	9	19.00	Silicon	Si	14	28.09
Francium	Fr	87	(223)	Silver	Ag	47	107.870
Gadolinium	Gd	64	157.25	Sodium	Na	11	22.9898
Gallium	Ga	31	69.72	Strontium	Sr	38	87.62
Germanium	Ge	32	72.59	Sulfur	S	16	32.064
Gold	Au	79	196.97	Tantalum	Ta	73	180.95
Hafnium	Hf	72	178.49	Technetium	Tc	43	(99)
Helium	He	2	4.003	Tellurium	Te	52	127.60
Holmium	Ho	67	164.93	Terbium	Tb	65	158.92
Hydrogen	H	1	1.0080	Thallium	Tl	81	204.37
Indium	In	49	114.82	Thorium	Th	90	232.04
Iodine	I	53	126.90	Thulium	Tm	69	168.93
Iridium	Ir	77	192.2	Tin	Sn	50	118.69
Iron	Fe	26	55.85	Titanium	Ti	22	47.90
Krypton	Kr	36	83.80	Tungsten	W	74	183.85
Lanthanum	La	57	138.91	Uranium	U	92	238.03
Lawrencium	Lw	103	(257)	Vanadium	V	23	50.94
Lead	Pb	82	207.19	Xenon	Xe	54	131.30
Lithium	Li	3	6.939	Ytterbium	Yb	70	173.04
Lutetium	Lu	71	174.97	Yttrium	Y	39	88.91
Magnesium	Mg	12	24.312	Zinc	Zn	30	65.37
Manganese	Mn	25	54.94	Zirconiuim	Zr	40	91.22
Mendelevium	Md	101	(256)				

Based on mass of C^{12} at 12.000. Values in parentheses represent the most stable known isotopes for elements that do not occur naturally.

Appendix C
Tabulation of Laser Data

MATERIAL (SYMBOL)	TEMP ° K	PUMP REGION (A)	WAVELENGTH (microns)	THRESHOLD (joules)
$CaWO_4;Nd^{3+}$	77	5700-6000	A 1.0650	1.50
			B 1.0633	14.00
			C 1.0660	6.00
			D 1.0576	80.00
			E 1.0641	7.00
	295		A 1.0652	3.00
			D 1.0582	2.00
$SrWO_4;Nd^{3+}$	77	5700-6000	A 1.0574	4.70
			B 1.0627	5.10
			C 1.0607	7.60
	295		B 1.0630	180.00

MATERIAL (SYMBOL)	TEMP °K	PUMP REGION (A)	WAVELENGTH (microns)	THRESHOLD (joules)
$SrMoO_4;Nd^{3+}$	77	5700-6000	A 1.0640	17.00
			B 1.0652	70.00
			C 1.0590	150.00
			D 1.0627	170.00
			E 1.0611	500.00
	295		A 1.0643	125.00
			F 1.0576	45.00
$CaMoO_4;Nd^{3+}$	77	5700-5900	1.0670	100.00
	295		1.0673	360.00
$PbMoO_4 Nd^{3+}$	295	5700-5900	1.0586	60.00
$CaF_2;Nd^{3+}$	77	7000-8000 5600-5800	1.0457	60.00
$SrF_2;Nd^{3+}$	77	7200-7500	1.0437	150.00
	295	7800-8100	1.0370	480.00
$BaF_2;Nd^{3+}$	77	5700-6000	1.0600	1600.00
$LaF_3;Nd^{3+}$	77	5000-6000	A 1.0631	93.00
			B 1.0399	75.00
	295		A 1.0633	150.00
$CaWO_4;Ho^{3+}$	77	4400-4600	2.0460	80.00
			2.0590	250.00
$CaF_2;Ho^{3+}$	77	4000-6600	2.0920	260.00
$CaWO_4;Tm^{3+}$	77	4600-4800 17000-180 00	1.9110 1.9160	60.00 73.00
$SrF_2;Tm^{3+}$	77		1.9720	1600.00
$CaF_2;Tm^{2+}$	20	2800-3400 3900-4600 5300-6300	1.1153	450.00
	77		1.1153	800.00

Appendix D
Laser Technical Data

ACTIVE MATERIAL AND VALANCE	OUTPUT WAVELENGTH	HOST MATERIAL	OPERATING MODE	OPERATING TEMPERATURE °C	
EUROPIUM (3+)	0.61	yttrium oxide plastic chelate in alcohol	PULSED	20	
CHROMIUM (3+)	0.70	aluminum oxide	CONTINUOUS	20	
SAMARIUM (2+)	0.71	fluorides of calcium strontium	PULSED	−196	SOLID OR IONIC LASERS
YTTERBIUM (3+)	1.02	glass	PULSED	−196	
PRASEODYNIUM (3+)	1.05	calcium tungstate	PULSED	−196	
NEODYMIUM (3+)	1.06	various fluorides, molybdates and glass	CONTINUOUS	20	

ACTIVE MATERIAL AND VALANCE	OUTPUT WAVELENGTH	HOST MATERIAL	OPERATING MODE	OPERATING TEMPERATURE °C	
THULIUM (2+)	1.12	calcium fluoride	PULSED	– 253	SOLID OR IONIC LASERS
ERBIUM (3+)	1.61	calcium-tungstate	PULSED	– 196	
THULIUM (3+)	1.91	cal-tungstate stron-fluoride	PULSED	– 196	
HOLMIUM (3+)	2.05	cal-fluoride cal-tungstate and glass	PULSED	– 196	
DYSPROSIUM (2+)	2.36 cal-fluoride		CONTINUOUS		GAS-DISCHARGE LASERS
URANIUM (3+)	2.4-2.6 poly-fluorides		CONTINUOUS		
HELIUM			CONTINUOUS		
NEON	160 wavelengths, between		CONTINUOUS		
KRYPTON	5,940 angstrom		CONTINUOUS		
XENON	units (0.594 micron) and 35		CONTINUOUS		
CARBON	microns				
MONOXIDE			CONTINUOUS		
OXYGEN			CONTINUOUS		
OTHER GASES			CONTINUOUS		
GALLIUM— ARSENIDE PHOSPHIDE	0.65-0.84		PULSED	– 175	SEMICONDUCTOR INJECTION LASERS
GALLIUM ARSENIDE	0.84		PULSED CONTINUOUS	20 196	
INDIUM PHOSPHIDE	0.91		PULSED CONTINUOUS	– 153 253	
INDIUM ARSENIDE	3.1		PULSED CONTINUOUS	– 196 296	

Appendix E
Laser Parts Suppliers

Beam Splitters
C.V.I. Laser Corp.
P.O. Box 11308
Albuquerque, NM 87112
Phone: (505) 296-9541

Coatings
Herron Optical, Div. Bausch & Lomb
2035 East 223rd Street
Long Beach, CA 90810
Phone: (213) 830-5404

Components
Coherent Radiation Co.
3210 Porter Drive
Palo Alto, CA 94304

Power Tech. Inc.
P.O. Box 4403
Little Rock, AR 72214
Phone: (501) 568-1195

Crystal Brewster Windows
Harshaw Chemical Co.
Broad St.
Gloucester, NJ 08030

Crystals
Adolf Meller Co.
P.O. Box 6001
Providence, RI 02940

Metrologic Instruments Inc.
143 Harding Avenue
Bellmar, NJ 08030
Phone: (609) 933-0100

Diode Pulsers
Power Tech. Inc.
P.O. Box 4403
Little Rock, AR 72214
Phone: (501) 568-1995

Dyes

ExCiton Company
5760 Burkhardt Rd.
Dayton, OH 45431
Phone: (513) 252-2989

Electronic Components

Havilton Electro-Sales
340 Middlefield Rd.
Mt. View, CA 94041

Mazda Electronics
1287 Lawrence Blvd.
Sunnyvale, CA 94086

Metrologic Instruments Inc.
143 Harding Avenue
Bellmar, NJ 08030
Phone: (609) 933-0100

Electronic Parts

Newark Electronics
500 N. Pulaski Rd.
Chicago, IL 60624

Radio Shack
2617 W. 7th Street
Ft. Worth, TX 79901

Electro-Optical Components

*RCA Radio Corporation
of America*
Electro-Optics Div.
Lancaster, PA 17604
Phone: (717) 397-7661
(RCA LED Diode,
#FLV-104 and S-6200
series diodes.)

Experimental Parts

Radio Shack
2617 W. 7th Street
Ft. Worth, TX 79901

Fabry-Perot Holders

C.V.I. Laser Corp.
P.O. Box 11308
Albuquerque, NM 87112
Phone (505) 296-9541

Fused Quartz Lens & Prisms

*Thermal American Fused
Quartz Co.*
Route 202 & Change Bridge Rd.
Montville, NJ 07045
Phone: (201) 334-7770

Gases

Matheson Gas Co.
Box 85
East Rutherford, NJ 07073
(Gases for science: CO_2, He,
neon, nitrogen, argon,
hydrogen, etc.)

Gas Lasers

*Hughes Aircraft Company,
Industrial Products Div.*
6155 El Camino Real
Carlsbad, CA 92008
Phone: (714) 438-9191

General Laser Parts

Cenco Scientific
2600 S. Kostner Ave.
Chicago, IL 60623
Phone: (312) 277-8300

Ealin Corp.
22 Pleasant St.
South Natick, MA 01760
Phone: (617) 655-7000

Edmond Scientific Co.
Dept. B-09, Edscorp Bldg.
Barrington, NJ 08007

Esco Products
Oak Ridge Rd.
Oak Ridge, NJ 07438

Fisher Scientific
711 Forbes Ave.
Pittsburgh, PA 15219

Information Unlimited
Box 716
Amherst, NH 03031
Phone: (603) 373-4730

Lasermetrics
Teaneck, NJ 07666
Phone: (201) 837-9090

Pioneer Industries
10-A Haughey St.
Nashua, NH 03060
Phone: (603) 882-7215

Sargent Welch
7300 N. Linden Ave.
Skokie, IL 60076
Phone: (312) 677-0600

Solaser
Box 1005
Claremont, CA 91711

Spectra-Physics
1250 West Middlefield Road
Mountain View, CA 94040

V.W.R. Scientific
P.O. Box 1050
Rochester, NY 14603
Phone: (212) 294-3000

Glass Products
Arthur H. Thomas Co.
3rd and Vine Street
Philadelphia, PA 19105

Plasma Scientific Co.
P.O. Box 801
Cucamonga, CA 91730

Information & Data
United Electronics Institute
3947 Park Dr.
Louisville, KY 40216

Lasers (Assembled)
Hughes Aircraft Company,
Industrial Products Div.
6155 El Camino Real
Carlsbad, CA 92008
Phone: (714) 438-9191

Metrologic Instruments Inc.
143 Harding Avenue
Bellmar, NJ 08030
Phone: (609) 933-0100

Lenses
Lambda Optics
Berkley, NJ 07922
Phone: (201) 464-5060

Mechanical Positioning Devices
Burleigh Instruments Inc.
100 Despatch Drive
Box 270
East Rochester,NY 14445
Phone: (716) 586-7930

Mirror Kits
C.V.I. Laser Corp.
P.O. Box 11308
Albuquerque, NM 87112
Phone: (505) 296-9541

Herron Optical, Div.
Bausch & Lomb
2035 East 223rd Street
Long Beach, CA 90810
Phone: (213) 830-5404

P.T.R. Optics
145 Newton Street
Waltham, MA 02154
Phone: (617) 891-6000

Optical & Lens Supplies
Herron Optical, Div.
Bausch & Lomb
2035 East 223rd Street
Long Beach, CA 90810
Phone: (213) 830-5404

Lambda Optics
Berkley, NJ 07922
Phone: (201) 464-5060

Power Supplies
Power Tech. Inc.
P.O. Box 4403
Little Rock, AR 72214
Phone: (501) 568-1995

Raytheon Company
28 Seyon Street
Waltham, MA 02154

Pyrex Parts
Dow Corning Co.
(Nationwide—consult your
telephone directory).

Q-Switches
Cleveland Crystals Inc.
19306 Redwood Ave.
Cleveland, OH 44110
Phone: (216) 486-6100

Ruby Rods
Adolf Meller Co.
P.O. Box 6001
Providence, RI 02940

Simulators
Power Tech. Inc.
P.O. Box 4403
Little Rock, AR 72214
Phone: (501) 568-1995

Special Components
Adolf Meller Co.
P.O. Box 6001
Providence, RI 02940

Systems
Power Tech. Inc.
P.O. Box 4403
Little Rock, AR 72214
Phone: (501) 568-1995

Tubes
Coherent Radiation Co.
3210 Porter Drive
Palo Alto, CA 94304

Raytheon Company
28 Seyon Street
Waltham, MA 02154

Wave Plates
C.V.I. Laser Corp.
P.O. Box 11308
Albuquerque,NM 87112
Phone: (505) 296-9541

Appendix F
Index of Refraction

The index of refraction (Table F-1) represents the speed of light through a substance compared to the speed of light through a vacuum.

Snell's law of refraction relates that: If *n* represents the index of refraction, *i* the angle of incidence and *r* the angle of refraction, then:

$$n = \frac{\text{sine } i}{\text{sine } r}$$

Where sine is the trigonometric sine of the angle (Fig. F-1).

Angle of Refraction:

$$\frac{\text{Sine } i}{A - C} = \frac{\text{Sine } r}{D - B}$$

Example of index of refraction:

$$\text{Index} = \frac{\text{speed of light (vacuum)}}{\text{speed of light (glass)}}$$

$$= 186,000 \text{ miles / second}$$

$$= 124,000 \text{ miles / second}$$

$$= 1.50$$

Table F-1. Index of Refraction for Various Materials.

Vacuum	1.000
Air at 0° C	1.00029
Air at 30° C	1.00026
Water at 50° C	1.330
Ice at 0° C	1.310
Carbon Tetrachloride	1.460
Diamond	2.470
Glass (Crown)	1.510
Glass (Flint)	1.710
Glycereine	1.470
Alcohol (Ethyl)	1.360
Benzene	1.500
Carbon Dioxide	1.00045
Quartz (Fused)	1.460

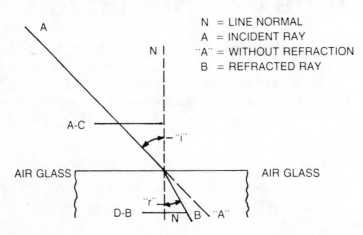

N = LINE NORMAL
A = INCIDENT RAY
"A" = WITHOUT REFRACTION
B = REFRACTED RAY

Fig. F-1. Angle of refraction.

Index

Other Bestsellers From TAB

☐ **LIGHT, LASERS AND OPTICS—John H. Mauldin**

A fascinating introduction to the science and technology of modern optics. Broad enough to appeal to the general science enthusiast, yet technically specific enough for the experienced electronics hobbyist, this book fully explains the science of optics. You'll explore: everyday observations on light, the theory and physics of light and atoms, computing with light, optical information storage, and many other related subjects! *Light, Lasers and Optics* is extremely well illustrated with over 200 line drawings. 240 pp., 205 illus.
Paper $17.95   **Hard $22.95**
Book No. 3038

☐ **FIBEROPTICS AND LASER HANDBOOK—2nd Ed.—Edward L. Safford, Jr. and John A. McCann**

Explore the dramatic impact that lasers and fiberoptics have on our daily lives—PLUS, exciting ideas for your own experiments! Now, with the help of experts Safford and McCann, you'll discover the most current concepts, practices, and applications of fiberoptics, lasers, and electromagnetic radiation technology. Included are terms and definitions, discussions of the types and operations of current systems, and amazingly simple experiments you can conduct! 240 pp., 108 illus.
Paper $19.95   **Hard $24.95**
Book No. 2981

Send $1 for the new TAB Catalog describing over 1300 titles currently in print and receive a coupon worth $1 off on your next purchase from TAB.

*Prices subject to change without notice.

To purchase these or any other books from TAB, visit your local bookstore, return this coupon, or call toll-free 1-800-233-1128 (In PA and AK call 1-717-794-2191).

Product No.	Hard or Paper	Title	Quantity	Price

☐ Check or money order enclosed made payable to TAB BOOKS Inc.

Charge my   ☐ VISA   ☐ MasterCard   ☐ American Express

Acct. No. _____ Exp. _____

Signature _____

Please Print

Name _____

Company _____

Address _____

City _____

State _____ Zip _____

Subtotal	
Postage/Handling ($5.00 outside U.S.A. and Canada)	$2.50
In PA add 6% sales tax	
TOTAL	

Mail coupon to:
TAB BOOKS Inc.
Blue Ridge Summit
PA 17294-0840   BC